Implication Analysis for Biotechnology Regulation and Management in Africa

Theorie in der Ökologie

Herausgegeben von Broder Breckling

Band 15

PETER LANG

Frankfurt am Main · Berlin · Bern · Bruxelles · New York · Oxford · Wien

Denis Worlanyo Aheto

Implication Analysis for Biotechnology Regulation and Management in Africa

Baseline Studies for Assessment of Potential Effects of Genetically Modified Maize (Zea mays L.) Cultivation in Ghanaian Agriculture

Bibliographic Information published by the Deutsche Nationalbibliothek
The Deutsche Nationalbibliothek lists this publication in the Deutsche Nationalbibliografie; detailed bibliographic data is available in the internet at <http://www.d-nb.de>.

Zugl.: Bremen, Univ., Diss., 2008

Typesetting & lay-out:
Richard Verhoeven

Figure Explanation:
ADINKRA symbols are ancient, traditional West African ideograms, see www.adinkra.org.
They are used today to express values, attitudes, and dedications.
For this publication we combined the symbol "Aya" (*Fern*, for endurance and resourcefulness) in the foreground with "Woforo Dua Pa A" (*when you climb a good tree*, symbol for support, co-operation and encouragement) on the background of a collection of maize cobs from different varieties and landraces.
Graphics by Denis Aheto and Broder Breckling.

D 46
ISSN 1615-374X
ISBN 978-3-631-59450-6

Printed in Germany 1 2 3 4 6 7

www.peterlang.de

Preface

The use of Genetically Modified Organisms - GMO - in agriculture is a globally controversial issue. Where they entered the market they led to disagreement. On the one side GMO are advertised as an important invaluable innovation, ***Profit, Progress, Future*** is promised. On the other side there are warnings, caution and criticism. ***Risk, Hazard, Damage*** are expected and analysed. Such a situation calls for independent information, and more: a well informed public sector, aware of who appropriates benefits, and who bears the risks. It requires well informed decision makers, who distinguish what is advertisement, single sided business communication, "public relation", PR and sales talk on the one side, and what is the entire range of implications and consequences – for those who grow the crops, who eat them and what the consequences for nature are – implications for the real world.

How is this for Africa?

African markets are likely to be confronted with GMO – but are the regulatory capacities prepared to investigate, check and assess the full spectrum of effects for African producers, for consumers, for the African environment on a comprehensive and advanced level? Actually, as the book confirms, there is an enormous lot to be done.

In the European Union as well as in the United States of America there are complex regulatory approaches and surveys established to defend the national interests and to balance the interest of agri-business companies, of the general public and the environment. Conditions in Africa are different, concerning societies as well as nature. Is as yet the *African* expertise in the position to take these differences into account, to foresee and prevent hazards in the best possible way when GMO admission is sought?

The work of Denis Aheto shows, that there are significant data and information gaps that can be closed - and he provides an important contribution. It is a key to strengthen the own position and not relying on foreign information but building on a domestic information basis. To deal with GMO is not just the adaptation of external views. In particular with regard to the agricultural structure and the characteristics of the environment, it can be detrimental to build decisions largely on external data, acquired on different continents. African agricultural practice and African environments are not the same as those for which agriculturally important GMO were developed.

In the current GMO controversy, only a very small part of research and public information deals with an independent analysis taking an African social and environmental perspective. The discussion about GMO in agriculture is largely dominated by foreign discourses and stakeholders. It is a frequent experience, that developments which are beneficial for certain stakeholders do not always increase the well-being of others, and in particular the larger parts of the rural population. In particular, there is a fundamental lack of environmental analysis.

As a matter of fact, it is fully undoubted that the vast majority of currently marketed GMO food crops do not bring advantage to the *consumer*. The argument in favour of GMO is a real (or imaginary) advantage for the producer and technology owner. Resistance genes against weed killing pesticides have been inserted into crop plants. This has been shown to increase the use of costly agro-chemicals. Or toxins against specific pest organisms have been inserted - bringing the problem that other, non-susceptible pests tend to come up. These two traits, herbicide resistance and insect pest resistance are currently the main transgenic traits world wide.

Genetic modification is not a kind of advanced breeding method. Plant breeding selects the most appropriate individuals. Genetic modification extracts Deoxyribonucleic Acid (DNA) as the carrier of genetic information from its natural context in one species and inserts it into a different species, usually involving additional artificial alterations that do not exist in nature, like combining viral and bacterial DNA to construct a transgene.

To assess not only apparent effects but also well as hidden results of genetic modification, a very thorough biosafety investigation is required.

Are GMO safe?
The relevant answer is
Depends what is meant by safe.

GMO are not comparable with other technical solutions. They require safety assessments profoundly different from the risk assessment of other goods. If a technical device turns out to be detrimental, it can be turned off - or disassembled. If a chemical that was released to the environment turns out to have unanticipated adverse effects, its release can be stopped. By the time it may dilute and degrade. Genetically modified plants can be pulled out at a particular site – but it is not always easy to eradicate them. They have the potential to self-reproduce, to disperse, to genetically re-combine and to self-modify along evolutionary traits. The genetic material is passed on to offspring, and as rare events, genetic material can even be horizontally transferred to bacteria and give rise to unexpected effects. In Europe, therefore, the establishment of a general surveillance for unforeseen GMO effects is a mandatory part of the regulation.

To make informed decisions about GMO - one must know a lot. In particular regulators have a high responsibility when they have to foresee, which the benefits are and which problems a GMO could bring to the country – bearing in mind potentially irreversible results of self-reproduction.

Agri-business is for profit. Therefore, it is consistent, that the first relevant applications are the ones that have the potential for the largest return to the developer - and this is applying the gene technologies to staple crops. Maize, soy beans, and wheat were primary targets. While maize and soy beans, predominantly grown in the South, entered the markets, GM wheat was not commercialised to-date. Mainly a severe resistance of

Canadian farmers blocked the introduction into the first large market and discouraged the patent holders from further steps with this crop species.

Safety is checked in the admission and notification procedure of each sovereign state. Testing is largely limited to short-term implications. What the long-term consequences for nutrition are - this remains still to a considerable extent untested.

Different strategies to safety assessment

It is important to note, that there are rather different regulatory regimes following different and to some extent incompatible strategies. One prototype of regulation is the system in the United States of America. Regulation there is fragmented between different authorities. EPA (Environmental Protection Agency), FDA (Food and Drug Administration), and APHIS (Animal and Plant Health Inspection Service) are involved for different GMO. Admission is granted comparatively easy, but if there should be harm after admision (to be proven in a lawsuit), there is a practically unlimited liability - the company that sold the GMO has to compensate for harm to others to unlimited extents. Regulation trusts, that market participants would successfully take legal action, if they feel their rights violated.

In the EU, a different approach was established. Admission is much more difficult to obtain. In a complex interaction of member states and EU institutions numerous details are checked before a notification is granted. The investigation beforehand aims to be comprehensive in order to avoid potential liability issues later.

There are aspects, that both regulatory approaches have in common. GMO notification is quite costly. Therefore, practically only "big players" can afford market access. In addition, none of the industrialised country allows GMO even as minor detectable impurities if they do not have a notification. There is a strict "zero tolerance" policy for GMO without approval. Harvests are not accepted for import when they contain any traces of GMO that are not notified. This is very relevant for smaller exporters when they have to bear additional cost of laboratory testing to prove that their harvests are free of particular genetically modified traits.

The contribution of this book

Are you concerned with the agricultural sector in the developing world? Do you have a stake in West African agriculture? Are you concerned about the prospects and fate of small holder farmers? Then this book offers relevant primary information.

It brings a good overview of the development background of GMO, and introduces into relevant dimensions of risk analysis. It provides the first genuine primary data for the West African context on the spatial structure of small holder farming in this context. It uses available information on cross-pollination distances in maize to predict the extent of crop impurities basing on a variety of different scenario assumptions. It expands on the socio-economic and gender implications and concludes, that only the very first steps in a West African Nation centered risk assessment have been made.

The author of this book took the opportunity to work in a large European project on biosafety research with more than 30 participating institutions. He contributed important field data - statistics how frequent some species would emerge outside cultivation or as an impurity in other crops. He extended the survey approach to maize crops to African conditions and developed an interdisciplinary layout. His data help to explain, which consequences an introduction of GMO into a smallholder farming context would have for crop purity issues, given the ownership, family, gender and subsistence situation of an African population majority. The book shows, that quite complex regulatory implications would arise that have to be well thought over in advance.

The findings are placed into a wider context of global trends. There is a strong support for the observation, that development generally is not a copy of others but requires the ability to defend an own view - as Immanuel Kant stated in his famous text on "What is enlightenment" of 1784: *sapere aude* – dare to know. Be couraged to use your own decision making potential without depending on the guidance of others.

The most important result of Denis Aheto's research: different from the situation in other countries that grow GM crops so far - in a smallholder context it would be much more difficult and much more costly to re-call GM varieties in a case of failures in GM crops. Therefore, regulation requires to operate on a "no second chance" base. Gene flow in the highly structured smallholder agriculture is considerably higher as in other agricultural contexts. No mistakes admissible, since no correction possible.

This is relevant not only for West Africa. It provides an example which is of equal relevance also for many other countries, which are confronted with world market relations and pressures to decide which options to take in agriculture. This book is a relevant contribution to

Problem formulation and option assessment,

clarifying who is the subject of development - encouraging to concentrate on an upgrading of traditional methods, and aiming at self-reliance and for exports at those premium markets delivering the highest return. For that, crop purity is a crucial condition.

In 2007, the German Academic Exchange Service awarded Denis Aheto, the author of this book at the University of Bremen for providing the most committed contribution combining scientific excellence, social awareness and mutual support.

Broder Breckling

Senior Lecturer for Ecology at the University of Bremen

There are known knowns, there are things we know we know
We also know there are known unknowns, that is to say
we know there are some things we do not know.

But there are also unknown unknowns. The ones we don't know we don't know, and if one looks throughout the history of our country and other free countries, it is the latter category that tends to be the difficult ones.

Donald Rumsfeld (2002), As an Expert of Non-knowledge

I dedicate this thesis to my dear family, Cynthia and Arnold Sefa Aheto,

and

beloved parents, Joseph and Fidelia Aheto

Contents

Part III: The Ghana case study on small-scale maize farming 63

Figures

Tables

Acronyms

AU	African Union
Bt	Bacillus thuringiensis
BMBF	Bundesministerium fuer Bildung und Forschung (German Federal Ministry for Education and Research)
BNARI	Biotechnology and Nuclear Agriculture Research Institute
CBD	Convention on Biological Diversity
CIMMYT	International Maize and Wheat Improvement Center
CRI	Crop Research Institute
CV	Commercial Variety
DNA	Deoxyribonucleic Acids
era	Environmental Risk Assessment
EFSA	European Food Safety Authority (EFSA).
EU	European Union
FAO	Food and Agriculture Organization
GAEC	Ghana Atomic Energy Commission
GEF	Global Environment Facility
GGDP	Ghana Grains Development Project
GIS	Geographic Information System
GLDB	Grains and Legumes Development Board, Ghana
GPS	Global Positioning System
GM	Genetically Modified
GMOs	Genetically Modified Organisms
GTZ	Deutsche Gesellschaft für Technische Zusammenarbeit (German Technical Cooperation)
Ha	Hectare (100 x 100 m)
Ht	Herbicide Tolerant
IITA	International Institute of Tropical Agriculture, Nigeria
ISAAA	International Service for the Acquisition of Agri-Biotech Applications
LMOs	Living Modified Organisms
MHs	Modern Hybrids
MOFA	Ministry of Food and Agriculture, Ghana
MON	Monsanto
OPVs	Open Pollinated Varieties
SIGMEA	Sustainable Introduction of Genetically Modified Crops into European Union Agriculture (EU Research Programme in the 6th Framework Programme)
TV	Traditional Variety (or Landraces)
UN	United Nations
UNEP	United Nations Environment Programme

Part I: Introduction

Chapter 1: The General context

1.1 Overview

Food security, loss of natural and agricultural diversity constitutes the greatest ecological challenge in human history (FAO, 2008). The way food is produced today, policies of industrial growth, agricultural trade and systems of industrial monoculture through the introduction of genetically modified (GM) crops pose immense implications on food sovereignty, crop purity and small farmers' livelihoods both in the north and south. In relation to developing countries, small- to medium-sized farms represent the most efficient source of food production when measured in terms of people fed per acre/ hectare of land (Planetdiversity, 2008). In West Africa, most farmers engage in various traditional and innovative forms of agriculture and organic food production, furthering genetic diversity and sustainable local economies (World Bank, 2005; Danso and Morgan, 1993). Organic food production requires registration, and the contracted obligation not to use specific inputs. This is not given in the smallholder context. However, a majority of farmers implicitly follow the rules of organic food production since the forbidden ingredients (synthetic agrochemicals and mineral fertilizers) are largely not used anyway. Food security is another key issue in its development, and as in the case of Ghana, many people depend on very few crops or eventually only a single crop- mainly maize to make the living (Morris, 1999). Therefore, maize crops are actively cultivated not only across the country but also across seasons and landraces represent an important genetic resource that deserves study and active conservation.

In the wider context of the global GMO debate, not only is maize one of the most relevant species in terms of its central role in nutrition, but it is also a high risk crop due to its extensive pollen dispersal features. Globally, it is also the species for which genetic modifications has reached the widest scale of commercialization. These issues have only served to bring more complexity in the broader GMO debate on risk implications. Hence GMOs are an issue. Their cultivation is a field of rapid development and considerable controversy worldwide. There are extensive economic interests behind it mostly driven by the biotech industry, there is consumer resistance largely experienced within Europe and outright rejection promoted by many social movements globally (e.g. see Greenpeace, 2007; FoEE, 2007). This is against the backdrop that GM contamination incidences have caused practically widespread economic and legal problems (Hewlett and Azeez, 2007). Now, largely remains on the level of advertisements as an option against hunger and to feed the poor countries in particular Africa. Presently, it is widely advertised in support of poverty alleviation and food security on the continent (ABSF, 2004; African Center for Biosafety, 2005). Within the African Union (AU), the question largely remains what the viable options are for a way forward? (Draft, African Model Law on Safety in Biotechnology, 2001).

Key issues relate to the use of genetic resources, related patenting (biopiracy) and systems of benefit sharing. Small indigenous farmers expect to benefit from their many years of sustaining seed biodiversity maintained over centuries. Also, the potential impact on local people especially women are discussed (e.g. deGrassi, 2003; Tsimese, 2003). Who should be held liable in case of adverse impacts due to GMO on local seed production systems, relevant redress procedures and systems of protection? In relation to this, the Cartagena Protocol on Biosafety since its inception in 2000 continues to negotiate a global regime of liability for adverse impacts caused by genetically modified organisms. While the Protocol has set minimum standards for import and export of GMOs, the industrialized countries including the United States pledged liability rules to take effect by 2007. This expectation met the contrary since insurance companies globally are not willing to cover possible damages from GMOs (see Planetdiversity, 2008). Despite the relevance of these factors, international discussions have focused largely on import and trade issues, industrial property rights, as well as agricultural intensification and have neglected specific points within the regional agricultural context.

In Ghana, several questions are posed among which include the current relevance of GMOs in Ghanaian agriculture, food safety issues and which available options could be harnessed to improve agricultural productivity (e.g. see GENET, 2005). However, a major problem relates to an absence of relevant data for risk estimation in basic crops particularly maize. This hinders reasonable judgements and obscures potential manageability implications. Once decisions are taken, and GM varieties are admitted, effects which may arise cannot be reversed. Hence, the need for a baseline anticipatory risk assessment carried out from a multidisciplinary perspective. As a preparatory phase to the study in Ghana, I participated for half year during 2005 (from May-October) in the Work Package Monitoring of the 6th EU Framework Project SIGMEA (Sustainable Introduction of Genetically Modified Crops into EU Agriculture, http://sigmea.go.dyndns.org/) that served as a methodological basis that was later adapted for the Ghanaian case. For data comparison with other project partners from Selommes France, Dundee Scotland and Mid-Jutland in Denmark, I gathered data on the population dispersal of Oilseed rape (*Brassica napus* L.) in rural agricultural sites of Bremen basing on aligned sampling protocol originally developed by Menzel in 2004. Further cooperation in the BMBF (German Federal Ministry for Education and Research) funded GeneRisk Project in the Social Ecology Call provided a maize model approach which has been applied to the Ghanaian situation to assess potential gene flow rates between GM and conventional fields.

Thus knowledge of the integrative approaches employed in the context of SIGMEA has been applied, however parametised under the local conditions e.g. climatic, socio-cultural and agro-structural factors. Therefore, for an overall estimation on the dispersal and persistence potential of transgene spread through pollen-mediated gene flow, we have inferred - in a separate chapter- to the recent study in 2005 on Oilseed rape (*Brassica napus*) crop production under European conditions. Thus, this thesis aims at covering a part of the analysis by assessing those factors, which may differ from other regions that could give rise to effects, which were not covered by previous risk analyses.

These factors cover in particular:

- Biotic (presence of organisms not relevant in other regions for which the GMO was tested, differences in receiving environments, and the physiological response of non-target species);
- Abiotic (referring to the pedological or climatic conditions),
- Molecular genetic factors (relating to genetic structure of landraces, addressing introgression issues, indigenous property rights, and potential consequences on native biodiversity);
- Agro-structural (differences in agricultural scale, crop management practice, cropping densities, spatial extents);
- Administrative (regulatory structure, estimation of efficacy of monitoring schemes and policy coherence); and
- Socio-economical (cultural backgrounds, crop management traditions, seed acquisition, selection criteria and preservation issues, and specific forms of local adaptation). Despite the relevance of these factors, they are largely ignored.

1.2 The development of transgenic crops

Until recent times, crop improvement largely resulted from conventional plant breeding involving the manipulation and recombination of genetic information within the same plant species. With the advent of genetic engineering, genetic information from other species can be incorporated into a crop genome. The development involves the introduction of new traits in crops that renders them insect resistant and/ or herbicide tolerant. This is achieved through the insertion of an artificial genetic construction, called transgene e.g. the insertion is done by particle bombardment frequently in monocotyles, while for dicotyles, insertion through *Agrobacterium* is efficient. These cells then regenerate new transformed plants generally referred to as Genetically Modified Organisms (GMOs) (Seralini, 2005). Insect-resistant crops that produce a protein which is toxic to some insect species and is coded by a transgene have been suggested to provide cost-effective options for growers, in that it takes away the worry of significant pest damage occurring and prevents economic losses (ISAAA, 2006; Evenson, 2003). Transgenic herbicide-tolerant crops on the other hand, are intended to facilitate the adoption of no or reduced tillage practices with resultant savings in time and equipment usage. Easier weed control measures may also reduce harvest costs, improve harvest quality leading to higher levels of quality price bonuses; especially for developing countries (ISAAA, 2006; Nuffield Council on Bioethics, 1999; ACDI/VOCA, 2003; BiotekAfrica News, 2003; ABSF News, 2004). Over here, GMOs are advertised as necessary in particular with the argument that it is needed to feed the poor countries. Maize is a major category of crops engineered to be insect-resistant or herbicide-tolerant, with the most relevant in this study being the former for which the modification leads to the expression of an insect toxic protein, originating from *Bacillus thuringiensis* (Bt).

Unlike chemical insecticides that are sprayed typically in a liquid form one or more times a season to control in particular Lepidopteran insects, Bt toxins are produced in the plant tissue continuously during the lifetime of the plant. The technology therefore is inherently prophylactic, since the plant produces the Bt toxins whether or not insect populations invade the plant or exceed economic thresholds (the level of pest infestation at which crop losses exceed the cost of control interventions). The other transgenic trait aiming at herbicide tolerance allows for the use of herbicides in all phases of plant development without damaging the crop (Benbrook, 2004). In general, alongside the rapid and widespread adoption of transgenic crop technology in some countries has emerged renewed interests concerning the potential risks posed by GM products, resulting from modern biotechnology.

1.3 Transgenic crop cultivation worldwide

The concept of transgenic crops was developed in the 70s, first produced and marketed in the 80s and cultivated on large-scale in the 90s (Traavik, 2008). It was first commercially introduced in China during the early 1990s primarily virus-resistant tobacco and tomato, and since then, the cultivation of transgenic crops has expanded rapidly worldwide. During 1995, numerous transgenic crops were commercialized, and by 1996, the USA was planting more transgenic crops than any other country in the world (Andow and Hilbeck, 2004). At present, the cultivation of transgenic crops is concentrated in four countries namely, USA, Canada, Argentina and China (ISAAA, 2006). Although they are not yet very widespread in Europe, they are currently grown in Spain, Germany, France, the Czech Republic and Portugal (Della Porta et al., 2006). They are also being grown in and targeted at more countries in the developing world as well (Andow and Hilbeck, 2004; ActionAid, 2003). Initial focus has been on four crops, namely soybean, maize (corn), cotton and oilseed rape (canola). Herbicide-tolerant soybeans, maize, cotton and oilseed rape, and insect-resistant maize, and cotton, account for over 99% of all commercial transgenic crops grown worldwide (Andow and Hilbeck, 2004; ISAAA, 2006). With the exception of cotton, these crops are mainly used for animal feed. Soya and vegetable oils derived from rapeseeds are used in processed foods (ActionAid, 2003). Transgenic maize and soybean varieties have also been utilized for the production of pharmaceutical drugs, vaccines, and other industrial chemicals (Association for Freedom of Choice and Correct Information, 2004).

Fig. 1.1 provides an overview of the total global plantings of transgenic crops till 2007. Fig. 1.2a shows an example of the main GM traits planted in 2005. GM Herbicide- tolerant (Ht) soybeans dominate accounting for 58% of the total followed by insect resistant (largely Bt) maize and cotton with respective shares of 16% and 8%. In total, herbicide tolerant crops account for 76%, while insect resistant crops account for 24% of global plantings. On the country level, the United States had the largest share of global GM crop plantings by 2005 (55%: 47.4 million ha), followed by Argentina (16.93 million ha: 19% of the global total). The other main countries planting GM crops in 2005 were Canada, Brazil and China (see Fig. 1.2b). Within the African context, South Africa constitutes the largest commercial GM crop producer, growing nearly 156,000 ha of GM Ht soybean, 385,000 ha Bt maize and 17,000 ha Bt cotton cultivars in 2005 alone (ISAAA, 2006). Within the East African region, Kenya for instance has started some field trials with GM maize, designed to control stem borers (Andow and Hilbeck, 2004) while countries such as Ghana in the western part, do not produce any biotech crops (African Center for Biosafety, 2005; GAIN, 2005). There are also no known GM field trials, contained or commercial releases to date in the country (GAIN, 2005; Ghanaweb, 2005). Nonetheless, GM crops are discussed as an option, in response to demands for appropriate measures needed for enhancing agricultural productivity in the country (Ghanaweb, 2005; SeedQuest, 2005).

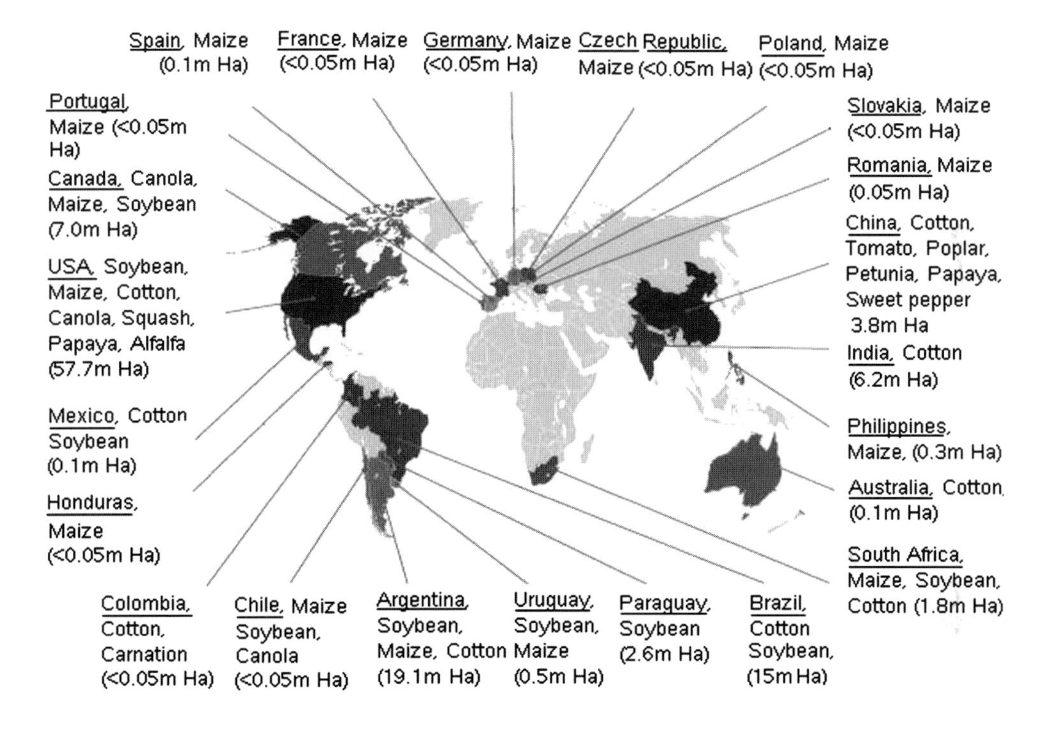

Figure 1.1: Biotech Crop Countries: Source: Clive James (2007), ISAAA (International Service for the Acquisition of Agri-biotech Applications. ISAAA (http://www.isaaa.org/default.asp) is a not-for-profit organization that serves as a mouth-piece of the biotech industries.

1.4 The aims of the Ghana case study

GMO crops were originally developed for rural large-scale agriculture mainly in developed countries, for which GM risk assessments have been primarily based. The aim of this project was to assess implied regional, ecological and socioeconomic impacts with respect to their potential introduction into subsistence small-scale farming in Ghana (e.g. Fig. 1.3). The introduction of new genes could lead to new ecological properties, eventually combinatory effects in the long-term. Pre-release monitoring is necessary to estimate dispersal processes for biosafety of the local population. Data acquisition for this study involved eleven months of fieldwork executed in 2006, and cooperation with the UNEP-GEF National Biosafety Project Secretariat in Accra. We investigated the situation for a 25 km^2 coherent area using specific habitat mapping procedures in (peri-) urban areas of Accra. Methods in the local context have been adapted from various data gathering attributes and in particular those employed in the 6th Framework EU Project SIGMEA (see Fig. 1.4), and familiarity gained with EFSA (European Food Safety Authority) status guidelines on risk analyses (EFSA, 2008).

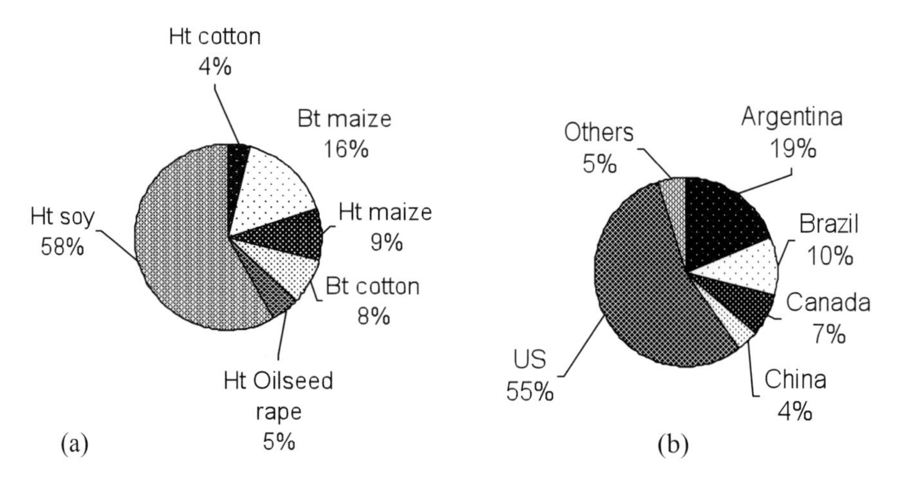

Figure 1.2: Global transgenic crop plantings by main trait and crop (a); and (b) by country in 2005, ISAAA (2006).

1.4.1 Study objectives

It was the objective of the study to:

- Acquire methods used for risk assessment within the international context, adapting them to the local Ghanaian situation. This objective therefore makes reference to the joint study within the European context on the population dispersal of oilseed rape, as an additional element of risk assessment of transgenic varieties (assessing hybridization probability and dispersal issues).
- Analyze the spatial context of maize cultivation practices in Accra, calculate farm isolation distances to estimate the feasibility of co-existence of GM and non-GM cropping measures.
- Estimate the degree of genetic variation of seed landraces in relation to commercially sown varieties. A first documentation of genetic variability of farmer samples is expected to play a key role in the further development of monitoring schemes due to the introduction of insect resistance.
- Investigate the synchrony effects of flowering states of crop and feral plants. This is expected to support an evaluation of self-dispersal characteristics of the different plant stands.

- Evaluate prevailing socio-economic implications of traditional farmer practices, seed acquisition criteria, and procurement sources. This objective estimates the extent of seed trait segregation and exchange among farmers.
- Assess the demographic structure of small farmers, their resource ownership, land rights, and gender effects. Furthermore, to relate these factors to their modes of farm production i.e. whether commercial or for subsistence. These are expected to provide an understanding on the extent of small-scale commerce and the level of dependency on harvests;
- Calculate cross fertilization rates between GM and conventional crops, and estimate the average GM input in conventional harvests.
- Assess the feasibility of 'coexistence' between GM and conventional maize production through ecological modeling.

1.4.2 Study hypotheses

The following hypothetical questions were analyzed in relation to gene-flow implications and general welfare factors for the peri-urban context:

- Cross field fertilization decreases with increasing field distance and increasing field size. Cross fertilization is higher between small and adjacent fields (This assumption is built into available pollen dispersal models however no parametization exists for tropical conditions and small-scale farming).
- The smaller the operated plots, the higher the heterogeneity of seed sources implicitly leading to increased genetic exchange and variability.
- There is significant phenological synchrony between cultivated maize crops and their wild feral populations.
- Seed reproduction from previous harvests does not differ significantly among subsistence and market-oriented farmers.
- There are significant gender differences in farm resource ownership, number of farms, and acreage, with women mostly in the informal sector growing maize mainly for subsistence.
- Male farmers and their female counterparts do not significantly invest in off-farm seeds and other productivity-enhancing technologies due to generally low income status. Hence, both groups show no significant differences in their seed acquisition arrangements and utility of modern varieties from the 'formal' seed system.

The thesis is divided into three parts. The first part is dedicated to the general introduction, objectives and background issues. The second part provides a chapter which discusses the Oilseed rape (OSR) study in Bremen within SIGMEA. The discussion of the results of this section is provided in the same chapter. The third part elaborates on the maize study in Ghana. The first part is divided into two chapters, the second part has one chapter, and the third part is subdivided into five chapters. In Chapter 2, the concept of risk assessment of GMOs is provided alongside relevant legal and policy issues. Chapter 3 presents the studies on OSR carried out in the international contexts, comparing the situation in different EU countries. In Chapter 4, an overview of maize reproductive biology, examples of specific transgenic maize events are provided. Furthermore, an analysis of local maize production in Ghana and relevance for biosafety monitoring are explained. Issues related to urban agriculture are then given. Chapter 5 outlines the methods of the study. In Chapter 6, the main results are presented, including outcomes of statistical evaluations conducted. Chapter 7 discusses the results of the study. Chapter 8 provides conclusions and proposals made for further research.

Figure 1.3: Urban maize farming in a suburb of Accra West, Ghana (2006).

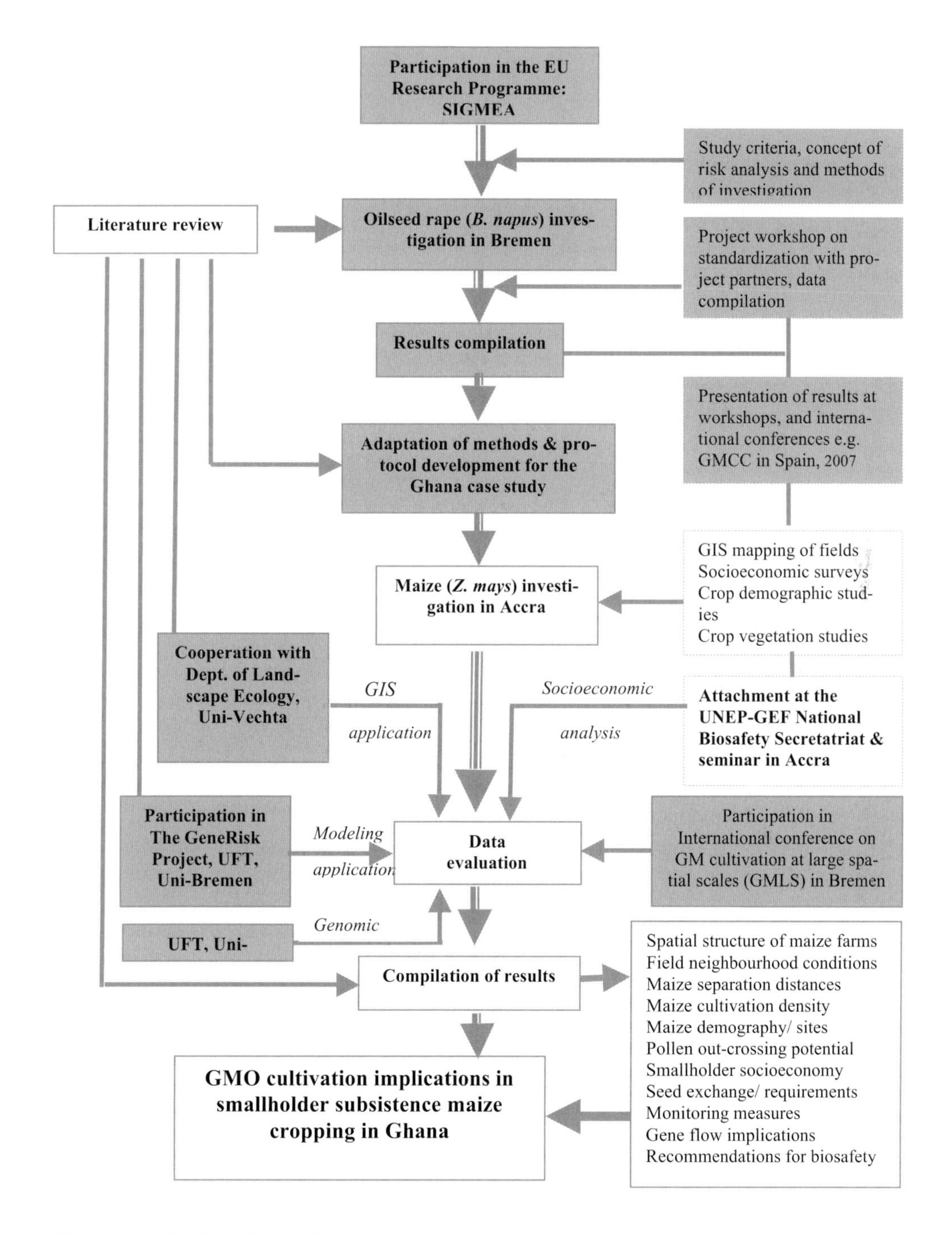

Figure 1.4: Outline of overall project approach. It focused on risk analysis of GMOs in small-scale farming in Accra basing on an exploratory study following participation in the interdisciplinary study among others, the 6th EU Framework Project SIGMEA (Sustainable Introduction of Genetically Modified Crops into European Agriculture), the GeneRisk Project, funded by BMBF (German Federal Ministry for Education and Research under the Social Ecology Call) and the GMLS (International Conference on the Implications of GMOs at Large Spatial Scales).

Chapter 2: Risk assessment of transgenic crops

Risk assessment is a systematic procedure for predicting potential risks to human health or the environment. In the context of transgenic crops, it involves the identification and evaluation of potential adverse effects of living genetically modified organisms on the conservation and sustainable use of biological diversity in the likely potential receiving environment, taking also into account risks to human health (Cartagena Protocol, 2000). Overall, it has been shown that gene flow in itself is not necessarily a risk but the effects that it may have. Gene flow from GM plants into domesticated or a wild population is ecologically relevant process that may shape the structure of genetic diversity with potential adverse effects on cultivated crops, wild and non-target species (Lecomte et al., 2007; Messěan and Angevin, 2007). Gene flow has also been discussed to have important operational, economic, and legal impacts for producers (Knispel et al. 2007). Risk assessment therefore considers the environmental and agronomic impacts of gene flow depending on the specific trait or plant combination and the likelihood of gene transfer (i.e. Risk = probability x extent of potential damage). Environmental risk assessment also considers the direct effects of introduced genes in crop management e.g. Bt effects on non-target organisms (e.g. see Andow and Hilbeck, 2004; and Heinemann, 2007).

2.1 Risk implication analyses in the context of transgenic crop cultivation

The cultivation of GMOs has raised concerns relating to environmental and health risks or unintended effects posed through potential genetic contamination by transgenes in managed non-transgenic conventional production fields (Pengue, 2004; Breckling and Menzel, 2004; ETC Group, 2003; Aylor, 2003a; Quist and Chapela, 2001; CEC, 2002). In the organic sector, the impact of GM contamination and the ability to continue farming organically is a major concern to their businesses. Organic food is enjoying increasing popularity and higher demand by the public due to its perceived higher quality (Hewlett and Azeez, 2007). There are concerns that GM pollen with insect resistance may pose potential hazard to non-target insect species (Aylor, 2002; Losey et al. 1999; Andow and Hilbeck, 2004). Even though some of these concerns have been refuted on technical grounds since the first scientific evidence of traditional maize crop contamination by DNA from transgenic maize in southern Mexico emerged (Christou, 2002), others have been heavily discussed. For example, GM contamination of traditional varieties of cotton in Greece (Kyriakidou, 2000), canola in Canada (Schmeiser, 1999), soy in Italy (Brough, 2001) as well as papaya in Hawaii (Greenpeace, 2003) have been reported.

Possible consequences of such contamination resulting from gene flow include genetic assimilation, wherein transgenes replace crop genes, and demographic swamping, wherein crop hybrids become less fertile than their parents, which implicitly leads to shrinkage of crop wild relative populations. Through these processes, there is also a possibility of fixation of disfavoured crop genes (Haygood et al., 2003). Transgenes also have the potential to confer on weeds a selective advantage, such as pest or herbicide resistance, which may lead to persistence or invasiveness of a species (Breckling and Menzel, 2004; Dale and Irwin, 1995), which could eventually lead to high costs of removal, if a necessity to remove transgenic individuals should arise from a future perspective (D'Hertefeldt et al., 2008; Breckling and Menzel, 2004). Effects on wildlife biodiversity, contamination of soil and water with broad-spectrum herbicides and other direct toxicological effects on the food chain have also been envisaged. From an agronomic standpoint, the transfer of novel genes from one crop to another could deplete the quality of conventional and organic crop seed, leading to a reduction in genetic diversity, changes in crop performance and lowering of the market value (Eastham and Sweet, 2002) thus presenting legal and trade repercussions (Aylor et al., 2003b). Changes in agronomic practices due to the introduction of GM crops could lead to efficiency problems in pest, disease and weed management (due to the increased use of herbicides and resistance development) (Breckling and Menzel, 2004). Seeds may be distributed through their dormancy mechanisms as well as in space (Eastham and Sweet, 2002).

2.2 Regulatory requirements of biosafety at the international level and sub-regional contexts

In the European Union and in the United States for instance, a precautionary approach to GM crops are taken, largely in response to farmers' and consumers' demands for biosafety in food and feed products (European Commission, 2007; European Union, 2007; FoEE, 2007). Risks are therefore assessed in advance and crops are authorized only if they are assumed not to pose additional risks to agriculture or health with respect to the varieties currently farmed. Concerning the quality of a GMO as food, the issue of substantial equivalence is assessed. In developing countries, the coming into force of the Cartagena Protocol on Biosafety in 2000 fulfills a similar obligation, by calling for scientific risk assessments of genetically modified organisms prior to their introduction into the environment. The Protocol addresses 'living modified organisms' (LMOs).

For the purposes of this thesis, the use of transgenic crops (GMO) refers to the definition of LMO in the protocol). As a protocol to the UN Convention on Biological Diversity, the Cartagena Protocol sets minimum standards for regulating certain aspects concerning the safe transfer, handling, and use of genetically modified organisms with special focus on the import and export of GMOs. In principle, these standards are expected to ensure an adequate level of protection to avoid or minimize potential adverse effects on the conservation and sustainable use of biological diversity, taking also into account risks to human health. In the African Union (AU), a strategy to implement the provisions of the Cartagena Protocol as well as address its gaps led to the development of the Draft Model Law on Safety in Biotechnology in 2001.

The activities which are not fully covered by the Protocol and required to be regulated additionally in national biosafety systems are:

- Development of domestic GMOs

- Use in contained systems, e.g. laboratories and production plants
- Approval of food consisting of or derived from GMOs.

The aim of the Model Law is to assist member states to develop comprehensive biosafety frameworks taking into account the sovereignty of states to regulate GMO issues but also their relevant international obligations. The challenges in implementing the protocol and the Draft Model Law led to the establishment of a Collaborative AU-German (GTZ) Project 'Capacity-Building for an Africa-wide Biosafety System', in view of the AU mission to produce guidance to harmonize political, legal and administrative procedures in the continent. The Model Law also obliges African states to develop risk management strategies that could help to protect human health and biodiversity, and to avoid any adverse consequences of unintentional releases of products of modern biotechnology (African Model Law on Safety in Biotechnology, 2001).

In Ghana, initial attempts of biotechnology regulation policy establishes a framework for developing systems to handle authorisations, including risk assessment, and monitoring of environmental effects, establishing also a need for public awareness and participation. The framework is in consonance with Ghana's constitutional obligations, environmental law and policy. The country is also Party to the Convention on Biological Diversity (CBD) and the Biosafety Protocol, both ratified on the August 29, 1994 and May 30, 2003 respectively. The CBD conferred on states the responsibility- including obligations- to formulate and implement strategies, plans and programmes for the conservation and sustainable use of biological diversity within their jurisdiction. Where necessary, to cooperate with other states in preserving biological diversity in areas outside national jurisdiction, taking precautionary, polluter-pays and preventive measures (Convention on Biological Diversity, 1992). These political requirements coupled with the civil concerns have posed a large demand for scientific baseline data allowing to estimate in advance potential risks as well as estimation of efficiency of monitoring measures within the West African region. However, it is important to note that the risks associated with a transgenic crop are related to the environment in which the crop is grown, with impacts varying according to how farmers perceive or take up the technology. It also depends on the scale and nature of the farming systems involved (Muhammed and Underwood, 2004).

Therefore, it can be concluded that risk assessment is a crucial requirement of the regulatory process of GMOs, and in Ghana, political will concerning the regulation and monitoring of GMOs exists, also in consonance with the African Model Law on Safety in Biotechnology as well as the relevant international regulatory requirements on biosafety. The next section focuses on aspects of field-based method development achieved through the involvement in international projects on risk assessment of GMOs.

Part II: Method development in the international context

The 6th EU Research Framework Project SIGMEA (Sustainable Introduction of Genetically Modified Crops in EU Agriculture)

Regional analysis of GMO cultivation implications in Northern Germany

Chapter 3: Methodological approaches adapted from the international contexts

In relation to GMOs, it has become increasingly relevant that the accuracy and specificity of methods of risk assessment should be appropriate, widely applied and tested. Reliable region-specific methods are needed to be able to characterize crop growing areas, receiving environments, agricultural scale including where wild relatives are located in relation to areas of (potential) GM cultivation. For West Africa, specific locally applicable methods are largely lacking. Therefore, participation in international projects such as the SIGMEA (Sustainable Introduction of Genetically Modified Crops into European Agriculture) and the GMLS (International Conference on the Implications of GM Crop Cultivation at Large Spatial Scales) yielded significant preparatory and analytical background to this thesis. This chapter therefore presents aspects of field-based studies used for predicting the implications of GMO cultivation based on spatial analysis of pollen-mediated gene flow. Lessons have been drawn from these for the emerging analysis of the Ghanaian case study.

3.1 Estimations of outcrossing probability between oilseed rape fields (and feral/ volunteer hybridization partners) using distribution frequency and neighbourhood distance analysis

3.1.1 Introduction – The 6th EU Framework Research Project SIGMEA (Sustainable Introduction of Genetically Modified Crops in EU Agriculture)

The Project SIGMEA was funded by the European Commission in 2004 as part of a broader strategy to create a science-based framework to inform decision-makers on co-existence and impacts of GM crops across Europe. Among others, to also identify relevant mitigation measures to minimise gene flow and mixing of GM and non-GM crops, developing sampling and evaluation schemes for risk analysis of GMOs in European cropping systems (http://sigmea.go.dyndns.org/). For the policy-makers, the key target was to have a science-based toolkit that could help them predict ecological and economic impacts, and to formulate cost-benefit analyses of GMO introductions. Through SIGMEA, the largest collection of data on geneflow and persistence in Europe has been organized. Therefore, I refer to the study in Bremen in 2005 for which I had gathered the data within the oilseed rape (Fig. 3.1) monitoring context of the project basing on a joint protocol. The approach was later in 2006 adapted for the Ghanaian case.

In this section, I document the methods employed in SIGMEA to examine the effects of neighbourhood distance frequencies. We did not measure cross-pollination directly. The analysis of neighbourhood distance frequencies is relevant because it is a factor influencing the opportunity for cross-pollination between oilseed rape crops (OSR) and their wild feral populations. Ferals are defined here as those crop-derived plants that exist in areas of land such as waysides, waste land and field margins that are outside the cultivated parts of fields, and as such are distinct from volunteers that inhabit fields. Ferals and volunteers are by definition of the same species as the crop and may be genetically the same if they arise from the same crop variety. They are classed differently because they are subject to different environments and management.

Gene flow in oilseed rape (*Brassica napus*): Gene flow from crops to their wild relatives is attracting attention due to the increasing commercialization of GM crops. OSR is identified to be among the list of GM crops to be commercialized within the EU agriculture. The crop has been recognized to have one of the highest potentials for gene flow via seed and pollen, and can hybridize with wild relative populations (Lecomte et al. 2007), capable of outcrossing, via potentially long-distance pollen movements (Devaux et al., 2007). Furthermore, seed escape from agricultural areas into urban settlement areas and near natural habitats has also been reported by Menzel in 2004. OSR volunteer populations also grow within crop fields (Lutman et al., 2005) or occur as feral populations within field margins, roadsides or occasionally on waste grounds (Pessel et al. 2001; Kawata, 2008). Subsequent introgression of crop genes could lead to the development of troublesome weeds and a niche expansion of introgressant populations (e.g. Ammitzbøll & Jørgenson 2005; Johannessen et al., 2006). High seed dormancy and persistence in seed banks of at least a decade as recently reported by D'Hertefeldt et al. (2008) may lead to volunteer populations that admix with and pollinate non-GM crops – and volunteers over time and spatial scales. For food safety standards, the question how many GM feral or volunteer plants could emerge in a field being grown with conventional oilseed rape after a GM crop? is crucial.

Figure 3.1: Oilseed rape field near Bremen (Northern Germany), 2005.

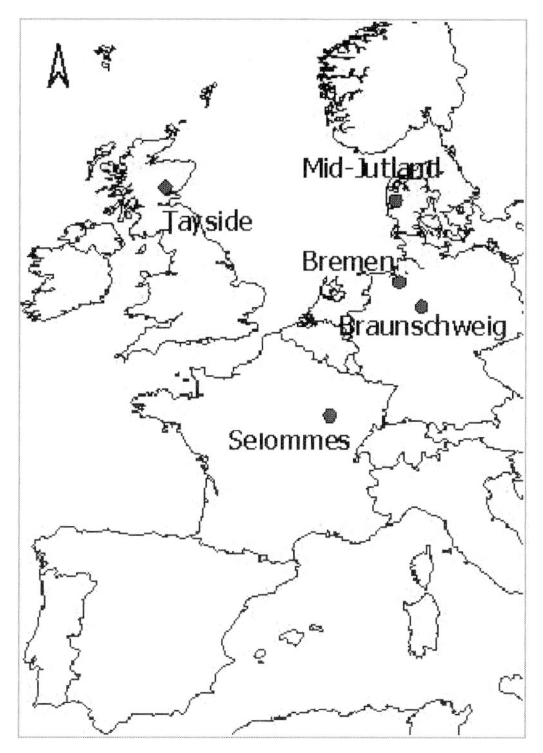

Figure 3.2: Locations of the five demographic study areas for feral oilseed rape, examined in SIGMEA made up of Tayside (UK), Mid-Jutland (Denmark), Selommes (France), Bremen and Braunschweig (Germany). Source: SIGMEA Final Project Report (2008).

The SIGMEA Joint OSR Study: For a European comparison on the risks implied in the cultivation of transgenic OSR crops, new research was initiated within SIGMEA as a key measure to compile data available for OSR across Europe; comparing data across spatial scales in the various regions; analysing hybridization probability between crops. Furthermore to identify specific knowledge gaps on gene flow to define further research. The joint tasks not only provided extended data on these dynamics but also analyzed the genetic composition, and related features of OSR crops. Five study areas were identified where populations of ferals in the landscape had been previously recorded along with volunteers and sown crops.

These covered areas in Tayside in UK, Bremen in Germany and Selommes in France, and additionally areas of Mid-Jutland in Denmark and Braunschweig in Germany (Fig. 3.2). Each region was represented by a 40 km^2 research area, defined by the location, year of study, typical crops, the area classed as arable land from government statistics (e.g. Table 3.1), and the fractional area of oilseed rape during the study estimated from GIS records. Initial studies in Bremen analysed data on origins, locations, persistence and potential for cross pollination (Menzel, 2006). Thus, SIGMEA offered a possibility to extend and assess other country-specific information available on these features before more general conclusions could be made on the European level.

The objectives: This paper focuses on the study in Bremen within the broader monitoring tasks to analyse the effects of isolation distances of OSR crops, feral and volunteers. The main objectives were to:

- Gather a data basis for frequency and distribution analysis of cultivated, feral and volunteer plants to enable a long-term monitoring strategy;
- Use GIS to characterise distributions and regions;
- Determine the potential for gene flow;
- Address inter-annual variability of OSR crop distribution in Bremen basing on previous studies; and
- Recommend coexistence management measures.

3.1.2 Methodology

A harmonized sampling protocol jointly developed by the partners was applied in data collection in a rural agricultural area of Hude 30 km west of Bremen (see Table 3.1 below). Subsequently, the data on the spatial orientation of fields, wayside ferals and wild relatives was systematized through a GIS database. Numbers of seed sown per unit area were also estimated as order-of-magnitude ranges. Flowering ferals as a percentage of all flowering oilseed rape was estimated as number of ferals (highest value in range) divided by the number of crop plants – the latter estimated as the product of GIS estimates of area grown with oilseed rape and typical crop stand density for the region. Cross pollination was not measured directly, but whether it occurs or not can be inferred from a more general knowledge of pollen transfer frequency with distance (e.g. Thompson, 1999). The field protocol sheet is similar to the one attached in Appendix 2.1. An overview of main data components is shown in Table 3.2.

Table 3.1: Characteristics of the rural study site in Bremen, Germany.

Item	Parameters
Latitude	58.05
Longitude	8.43
Area studied (km^2)	40
Year of measurements	2005
Main arable crops	Silo corn, barley, rye, winter wheat, winter OSR, triticale
Area of arable land as % of total land area	50
Season of sowing OSR crop: spring autumn (winter)	 4% 96%

Figure 3.3: Mapping of OSR (shown in circle) at a rural agricultural site in 2005 using a hand-held Garmin Etrex GPS. The photo depicts a feral location at the edge of a newly sown grain field in Hude (Gauss-Krüger coordinate: 3460502, 5879946).

Table 3.2: Field protocol and parametization.

Variable	Parameters and measured variables
1) Species identification	16 different species
2) Location types	30 different site categories
3) Plant frequency	Individual numbers (incl. height estimation and area coverage)
4) Phenological states	Emergence of buds, inflorescence etc. incl. vitality
5) Track point	Location/way point (Garmin eTrex GPS)
6) Sample collection	a) Rural/ agricultural area (Hude): 1st phase (May, 2005) - flowering phase, occurrence. - Leaf tissue sample collection - (2 leaves/ plant collected) 2nd phase (July-Sept, 2005) - reproductive success, mortality, - validation of GPS recordings - (collection of seed pods) b) Urban site (Bremen): 1st phase (June, 2005) - Leaf tissue sample collection 2nd phase (August, 2005) - Seed samples collection
7) Picture documentation and coding	Two or more pictures of plant and site
8) Leaf sample preparation & storage (-80˚C in refrigerator)/ seeds further dried at room temperature	– later: analysis of population genetic heterogeneity

(Source: Adapted from Menzel, 2006)

3.1.3 Results and discussion

Species looked for, species found: Species looked for, and species found in the rural sites in relation to those of urban sites are shown in Table 3.3. The considered species are closely related to oilseed rape and are considered as potential hybridization partners. More species were found in the urban sites, occurring mostly as wild ferals (e.g. no fields and volunteers) as compared to the rural sites, even though with lower number of species found occur as fields, wild ferals and volunteers.

OSR fields and wild relative distribution: Fig. 3.4 below show locations where populations of feral oilseed rape in the landscape had been identified and recorded along with volunteers, wild relatives (e.g. Fig. 3.6) and sown crops. Around 50% of the land surface is arable. An estimated 5 to 9% of the arable land is sown with oilseed rape, with most of them consisting of winter varieties. Feral crops constitute a very small fraction – less than 0.0001% - of the total oilseed rape plants in the region. This latter figure is estimated from the number of flowering and seeding ferals measured in the surveys as a percentage of the estimated total number of crop plants (field-areas determined by GIS). Also estimated was the number of cultivation sites, field sizes, number of feral and volunteer locations.

Table 3.3: Species looked for, species found in 2005.

	Species looked for	Species found	
		Rural sites	Urban sites
1	*Brassica napus*	O	O
2	*Brassica elongata*		O
3	*Brassica nigra*		O
4	*Brassica oleracea*		O
5	*Brassica rapa*	O	O
6	*Descurainia sophia*		
7	*Diplotaxis muralis*		O
8	*Diplotaxis tenuifolia*		O
9	*Diplotaxis viminea*		
10	*Erucastrum gallicum*		
11	*Hirschfeldia incana*		
12	*Raphanus raphanistrum*		O
13	*Raphanus sativus*	O	O
14	*Rapistrum rugosum*		O
15	*Sinapis alba*	O	O
16	*Sinapis arvensis*	O	O
Total number of species found		**5**	**12**

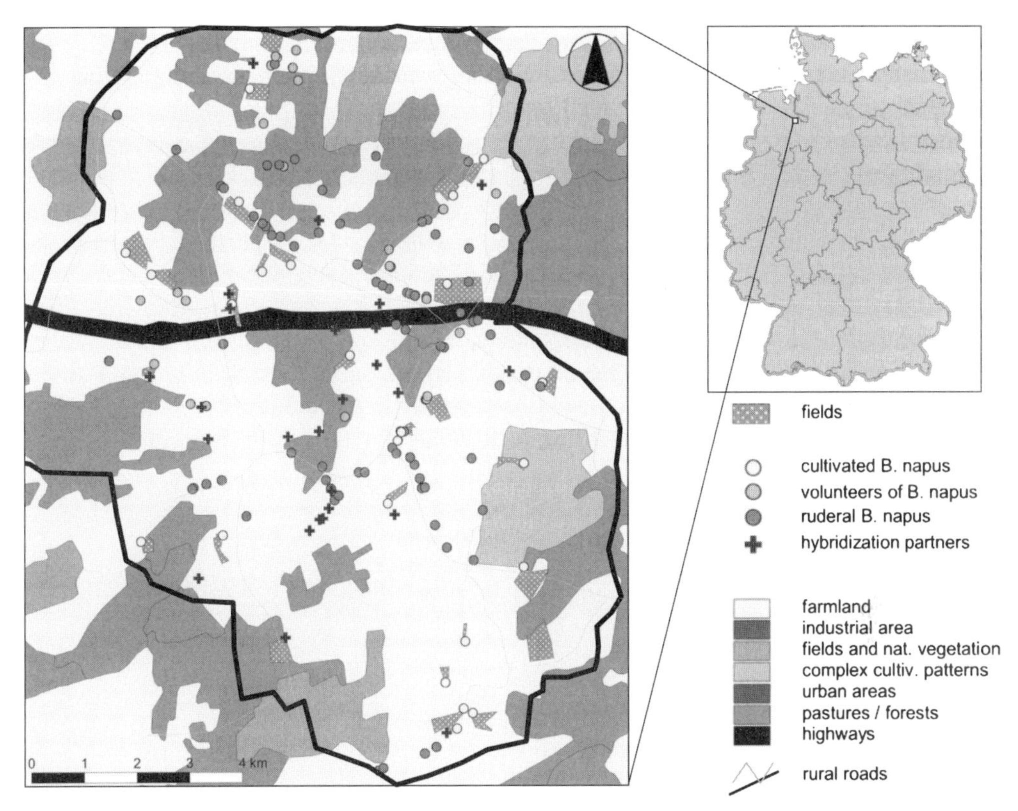

Figure 3.4: Distribution of oilseed rape fields and potential hybridization partners within an area of 40km2 rural agricultural site of Bremen, 2005. The data is presented on the background of a CORINE land cover map. Area of arable land as % of total land area = 50; Percentage of total land area sown with OSR during study = 2.5; Percentage of arable land area = 5.0.

Isolation distances of plant locations: In Fig. 3.5, an analysis of nearest distance relations between OSR feral stands (%) could be seen for the overall area of 40 km^2. Similarly, Fig. 3.7 presents the results of nearest distances between crop fields. The most important question relevant to coexistence in space is whether ferals can contribute genes to crops at a level that would raise GM impurities. The analysis is shown in Fig. 3.9. The data reveal a hyperbolic relation between the numbers of ferals and distance to nearest crop, with 40%, 23%, 14%, and 8% of ferals occurring within 100 m, 100-200 m, 200-300 m, and 300-400 m from crop fields. In general, all feral populations were found in the range of 1,100 m from OSR fields.

Site characterization and population densities: The analyses here relate to most small populations of oilseed rape that habitually occur along roadsides, field margins (Fig. 3.8) and waste ground (see Table 3.4). The most common population size was between 1 and 10 plants, but occasionally, areas of waste ground and construction sites supported much larger numbers. (When such large populations occur in derelict fields, they are classed as volunteers rather than ferals). Characteristics of the feral populations (Fig. 3.10) in the study area show certain key features:

- their main habitat is waysides; a secondary habitat is field margins in some areas,
- ferals are subject to weed control and mowing (variously 25 to 95% populations affected),
- the 'density' of populations is typically around 1 per km^2

However, demographic studies generally show that a small percentage of populations re-occur at the same site over successive years (Menzel, 2006), whereas a majority appear to die out rapidly (Dietz-Pfeilstetter et al. 2006; DETR 1999; Pivard 2006). It is problematic to assign origins to populations (e.g. from a nearby crop, from transport, from the seedbank), but evidence of persistence for several years at high abundance in the seedbank has been found in some areas (DETR 1999; Pivard et al. in press). A field experiment gave evidence that feral seeds can be dispersed along roadsides by turbulence associated with vehicles, but dispersal distances by this means were a maximum of 21 m (Garnier et al. in press).

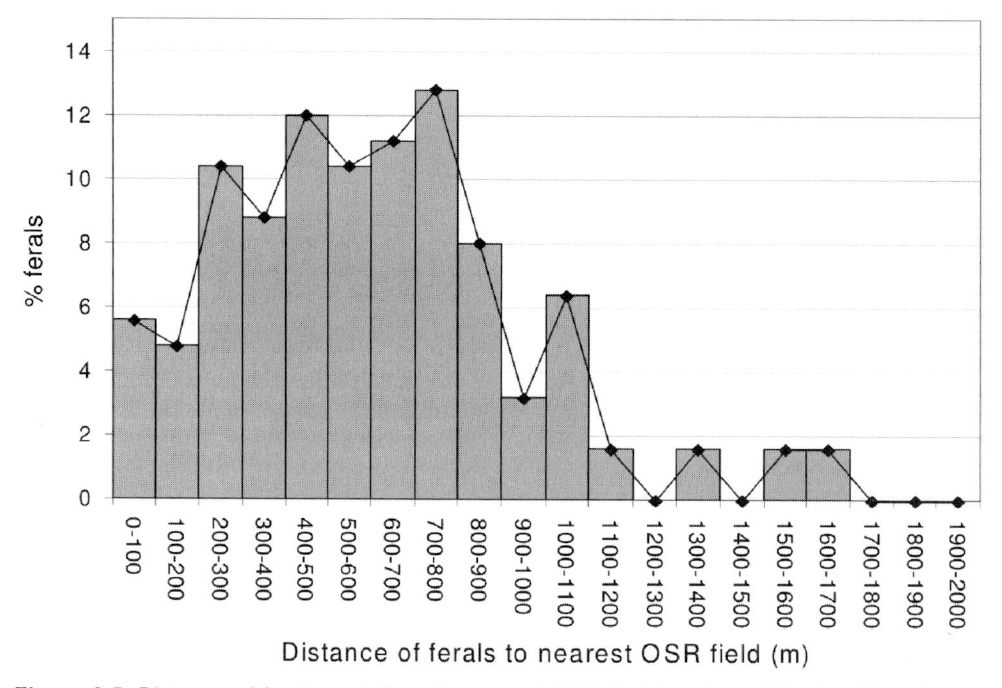

Figure 3.5: Distance of feral populations to nearest OSR field location within the 40 km2 area. The analysis shows that about 5.7% of the feral population are between 0 and 100 m away from the next field with OSR cultivation, 5% occur in a distance between 100 and 200 m, and around 10.7% between 200 and 300 m, etc. The largest observed distance between a feral occurrence and an OSR field was between 1600 and 1700 m.

Figure 3.6: Examples of wild relatives of OSR found. (a) *Raphanus sativus* (radish); (b) *Sinapis arvensis* (wild turnip); (c) *Sinapis alba* (white mustard); (d) *Brassica rapa* (field mustard).

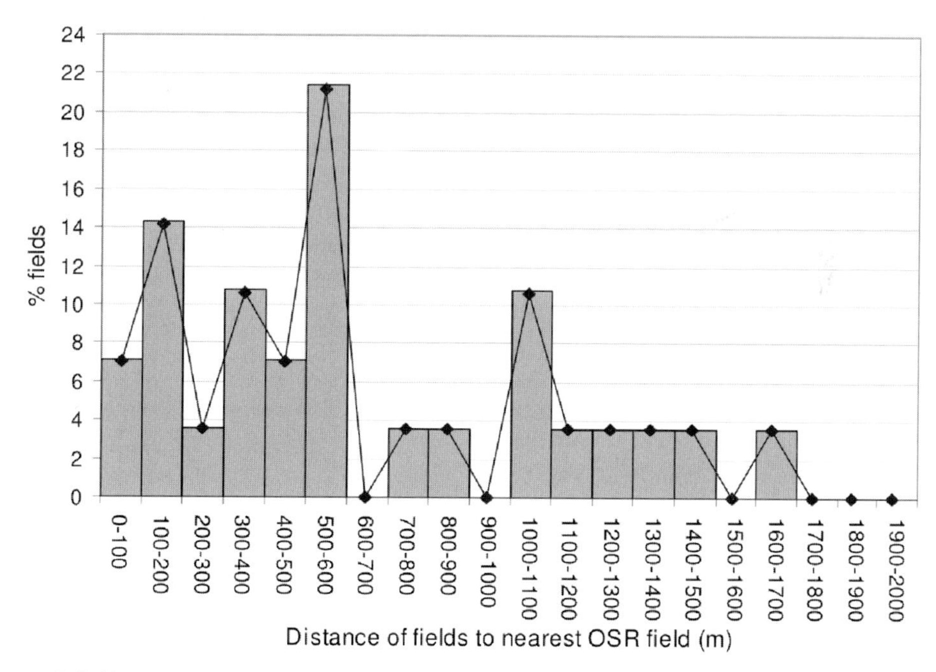

Figure 3.7: Nearest distances between OSR field locations within the investigated 40 km2 area. The distribution generally shows an irregular pattern. The analysis shows that about 7.0% of OSR field population occur between 100 m from the next field of OSR cultivation, 21% occur within 200 m and 25% of all fields are observed within a distance of 300 m. It is also observed that nearly 50% of all fields occur within 600 m of a field location, and the entire population of fields (100%) occurred within a distance of 1,700 m.

Figure 3.8: Feral plants along grain fields nearby an OSR field, and occuring between wheat field and pasture (Gauss-Krüger coordinate: 3461562, 5884033; Hude, May 2005).

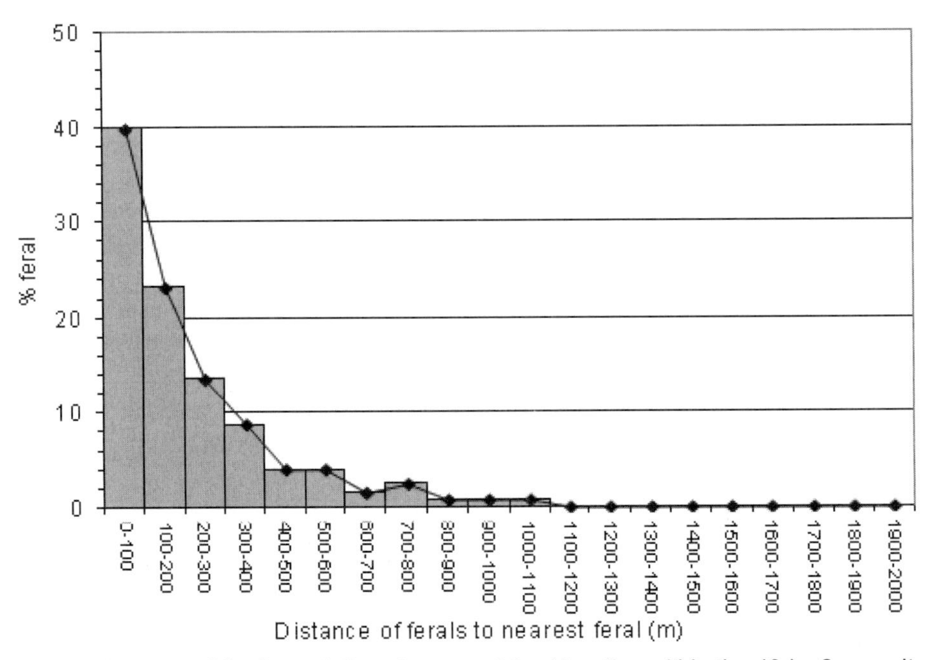

Figure 3.9: Distance of feral populations to nearest feral location within the 40 km2 area. It reveals that 40%, 63% and 77% of all ferals were found within nearest cumulative distance of 100m, 200m and 300m of a feral location. In principle, over 50% of all feral locations occurred within nearest distance of 200m of ferals. All ferals (100%) were located within nearest distance of 1,100 m.

Seeding rate/ reproductive success: Estimations of the number of flowering and seeding ferals were measured in the survey as a percentage of the estimated total number of crop plants (field-areas determined by GIS and using the typical or recommended crop density for the region). the number of seeding plants per km^2 was typically between 1 and 10, at least 80% of feral populations flowered at the same time as neighbouring crops. Ferals typically produced 100-1000 seeds per km^2.

Comparison of the Bremen data with results from other SIGMEA partners: The comparison of the obtained data with the results from other SIGMEA partners in different countries substantiated, that the results as presented above are not exceptional but rather typical for other sites in Europe. The proximity of ferals to the nearest flowering crop was measured in four of the study areas. Around 10 % of feral populations were within 10 m, 15% within 100 m (50% at Selommes) and 80% within 1000 m (Fig. 3.12). Cross pollination is most likely to occur frequently between crops and the 10% of ferals that lie very close to fields. The study did not examine the effect of crop- to feral hybridisation on the fitness of the individuals and the population.

Table 3.4: Characteristics of the feral/ volunteer populations in specific habitat types in the 40 km^2 agricultural area.

Wild plant demography	Feral:			Volunteers:		Total
	roadsides/ traffic areas	field margins	farm yard, garden, soil dump	within grain fields	within pasture/ meadows	
No. of feral found	74.0	18.0	5.0	-	-	97.0
% feral population	76.4	18.6	5.2	-	-	100.0
Density of Feral (km^{-2})	1.85	0.5	0.1	-	-	-
No. of volunteers found	-	-	-	24.0	8	32.0
% volunteer population	-	-	-	75.0	25	100.0
Density of volunteers (km^{-2})		-	-	0.6	0.2	
Mean density (km^{-2}) =	0.81			0.4		1.2
Density range (km^{-2})	0.45-1.85			0.2- 0.6		-
Total No. of feral and volunteers =						129.0

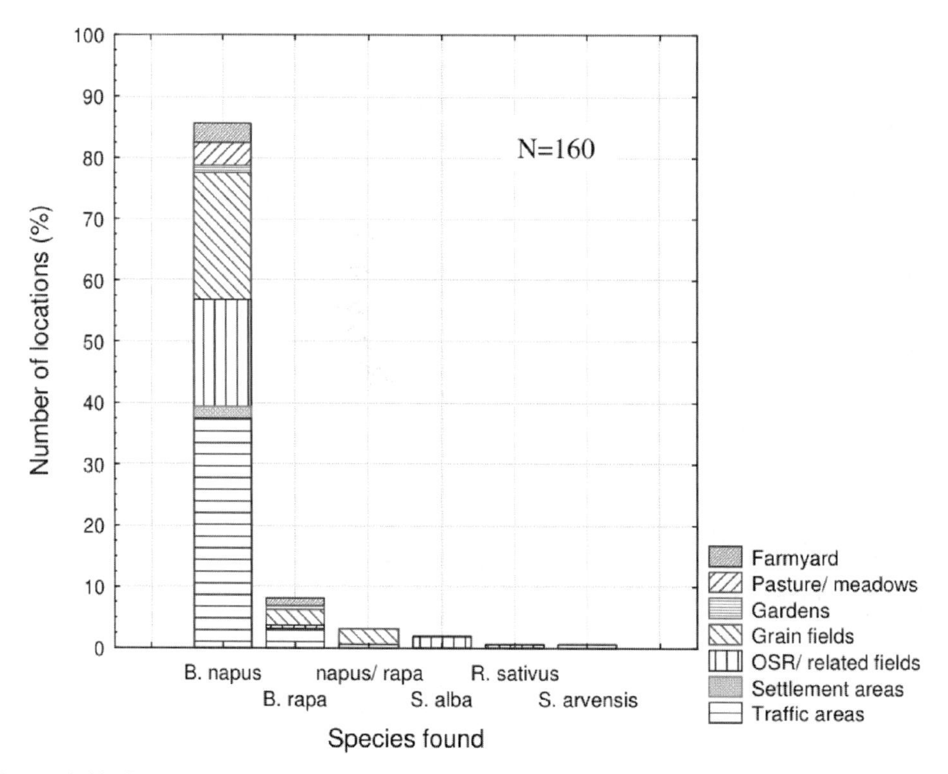

Figure 3.10: Demographic and site characteristics of oilseed rape and potential hybridization partners found in rural areas of Bremen. The designation napus/ rapa refer to situations where a species found could not be distinguished distinctly as either B. napus or B. rapa, and observed to be either one of them. B. napus occurred in the highest number of locations representing 85% of all locations found. This was followed by B. rapa occurring in the range of 8% of all locations. S. arvensis occurred in the least representing 0.4% of all locations. In terms of sites, majority of all locations were found in traffic areas, followed by grain fields, and gardens. The least were found in settlement areas.

Figure 3.11: Feral Oilseed rape (a) Feral OSR on farmyard manure (Location coordinate: 3460911, 5880664); (b) OSR field (Location coordinate: 3464173, 5882885); (c) Feral OSR in traffic areas (Gauss-Krüger coordinate: 3463030, 5879408); (d) Volunteer OSR in grain field (Gauss-Krüger coordinate: 3464256, 5880767).

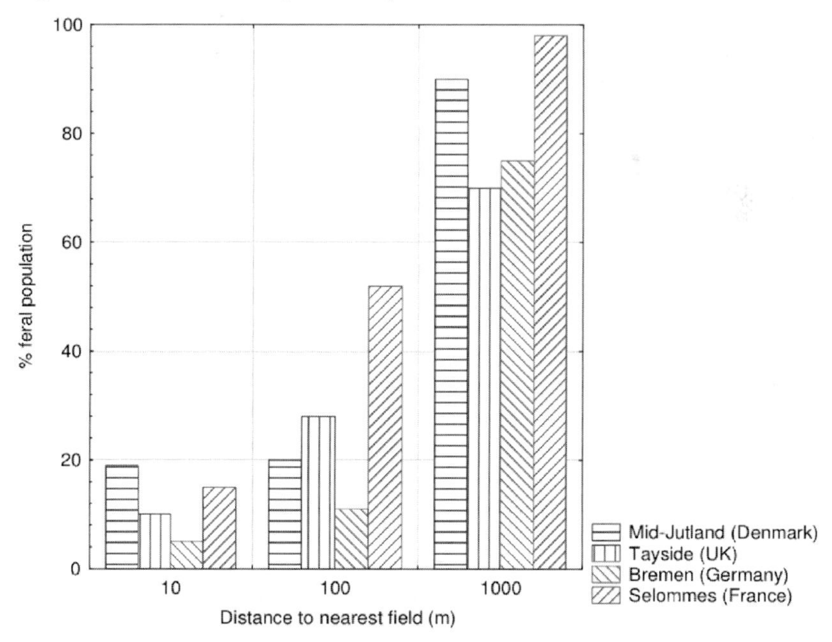

Figure 3.12: Nearest distance analysis between crops and ferals among the cooperation states. Result of a joint contribution by the partners as presented by Lecomte on behalf of consortium members at the GMCC (International Conference on the Coexistence of GM and Non-GM Crops) in Seville, Spain in November 2007. The Bremen data supported the overall trend observed in the other countries.

Estimations of gene flow between crop and ferals: The attributes of gene flow are shown in Table 3.5. Features of feral 'populations', defined earlier, and as measured in the field surveys show values determined as 'total number of distinct populations / total area of survey' and averaged over the years of the study, together with the highest and lowest values in any year. Estimates of population size were not entirely consistent between regions and are simplified here in terms of order-of-magnitude ranges for the number of flowering plants per unit area. This indicator is given for all populations, which includes some massive populations in derelict fields for example, and for typical wayside ferals which tend to be found in groups of less than 100 plants. Origins of the populations were estimated in several ways, in some instances from molecular typing and in other from direct observation.

Table 3.5: Assessing implications for gene flow between crop and feral plants in Bremen.

S/n	Descriptor	Result
1	Feral flowering simultaneously with crop (%)	80%
2	Feral seed km^{-2}	1000-10000
3	Feral plants as % of total oilseed rape flowering seeding	 <0.001% <0.0001%
4	Potential for seed shed from feral to germinate and grow into mature plants	yes
5	*Evidence of old varieties at feral sites	yes
6	% of feral populations within distance to nearest crop in flower 10 m 100 m 1 km 2 km	 5 10 80 -
7	Likelihood of gene flow from crops to feral	likely but very low
8	Feral population density (km^{-2}), mean and range	0.81 0.45-1.85
9	Total number of feral plants flowering (km^{-2}) all populations population up to 100 plants	 10-100 1-10
10	Habitats (%) of feral on roadside field margin other (e.g. soil dump, farmyard, garden)	 76 19 5
11	Possible origins vehicular transport seed bank local recruitment (e.g. shatter from fields)	 yes yes yes
12	% populations affected by weed management	60%

Site information gathered by Aheto, D. W. from May-August 2005, SIGMEA WP2 Monitoring of OSR crops in Bremen. * Evidence of old varieties at feral sites related findings of this study to those of Menzel in 2002 and 2003 (See Menzel, 2006). From the table, it could be derived that an efficient control of transgenes appears difficult.

3.1.4 Conclusions

Ecological and management implications: Even though feral populations occur at rather low densities biotic interactions between feral and crop plants co-occur and the efficient control of transgene dispersal seems highly unlikely. Volunteer dynamics are influenced by several factors mainly agricultural practices, potential utility of same farm machinery by different farmers and high seed dormancy considerations. Feral oilseed rape occurs widely on waysides, field margins and waste ground in the study area at a very low frequency of around 1 population per square kilometer; a typical population consists of 1 to 10 plants. This means that if GM varieties were grown widely, they would become constituents of feral populations. A 'worse-case' (a very highly improbable one) can be estimated based on the following: if it is assumed that all the feral plants in a region were GM and cross pollinated with conventional oilseed rape crops (equally plant per plant), then the level of impurity introduced from this source would be between 0.001% and 0.0001%.

At their present density and spatial arrangement, oilseed rape feral plants need not be considered in management prescriptions to achieve GM levels below 0.9% in non-GM crops, above of which harvests have to be labeled as genetically modified in the EU. Despite their low frequency of occurrence compared to volunteers, they have become one of the most common cruciferous plants of waysides and margins, and in some areas occur in mixed populations with cruciferous wild relatives. It is very likely that the introduction of GM oilseed rape in agriculture would lead to a dispersal of transgenes outside cultivation, where it is likely to persist for longer time spans.

The potential to introduce GM impurities: The probability to introduce transgenes through cross-pollination between crops is:

- Moderate for cross pollination between fields, however could be managed through spatial separation e.g. use of buffers where crops are in close proximity;
- Moderate through wild relatives in localised areas where they occur in high abundance in grain fields;
- Potentially significant due to long distance movements by insects (e.g. honeybees interaction range up to 5000 m) in the landscape may lead to background levels to augment impurities from other sources;
- Low during the first few years of commercialisation by management of cross pollination (through separation, etc.) and seed purity, but large uncertainties remain with regard to cumulative or combinatory effects in relation to the spatial and potentially long range movement of pollen when several different GM crops are admitted for cultivation.

Basing on the above, it can be concluded that the persistence of transgenes due to the spatial distribution of feral populations in relation to sown crops, gene exchange by pollen and seeds throughout the landscape limits the feasibility of defining simple rules for isolation between GM and non-GM co-existence.

3.2 Analysis of the spatial density and neighbourhood distances of cultivated oilseed rape (Brassica napus) fields in Northern Germany

Analyses of the relations in the cropping system should not be restricted to small, local scale. It is also of interest, which implications the geographic conditions of crop growing yield across larger regions. This is a field where only very few studies have been executed. As an example how larger areas can be assessed, the density of oilseed rape cultivation in Northern Germany was investigated using remote sensing as a data source to assess large-scale neighbourhood conditions.

3.2.1 Introduction

For the Northern German Federal States of Lower Saxony, Schleswig-Holstein, Mecklenburg Western-Pomerania and Brandenburg, the location and size of 38,625 oilseed rape fields was identified using remote sensing data (Laue, 2004). This allowed to assess neighbourhood conditions, which are crucially relevant to estimate the implications of large-scale commercial GMO cultivation and to address the feasibility and efficiency of coexistence measures, e.g. co-ordination efforts in the application of isolation distances.

Remote sensing area coverage with cloud-free images was best for 2001 to allow further analyses. Relevant parameter described for example by the mean field sizes, mean nearest distances between fields, number of field neighbours within specific distance ranges could help to quantify the potential of gene flow from GM to non-GM fields or potential interbreeding partners in the different regions.

3.2.2 Methodology

Overlapping LANDSAT 7 ETM+ remote sensing images were obtained through the distributor Eurimage (http://www.eurimage.com/) and used to identify agricultural oilseed rape fields (Fig 3.13). Separation of oilseed rape from other crops is best during flowering. Laue (2004) developed an algorithmic approach to optimise the distinction of oilseed rape from other crops. With a frame size of 183 km and a pixel size of 30x30 m, it was possible to locate fields down to a size of about half a hectare and obtain spatial coverage of most of the region under consideration. The field size and field centroid coordinates were used for further calculation of mean nearest neighbour distances and the frequency how often a particular number of neighbours occurred within a certain distance range.

3.2.3 Results

The neighbourhood conditions for the federal states analyzed yielded a characteristic local variation. Table 3.6 lists the number of field cases, the area of the federal state and the average distance of nearest neighbours. The data revealed the highest cultivation density in Schleswig-Holstein with the least recorded mean distance between nearest neighbours. Thus, the field centroid data confirmed a larger number of smaller fields in Schleswig-Holstein compared to the other Northern German federal states. The region of Brandenburg revealed the least cultivation density and the longest mean distance between nearest field locations. In Figs 3.14 A-D, it is shown in detail, how many neighbours are found within distances between 250 m up to 5000 m basing on the field centroid distances.

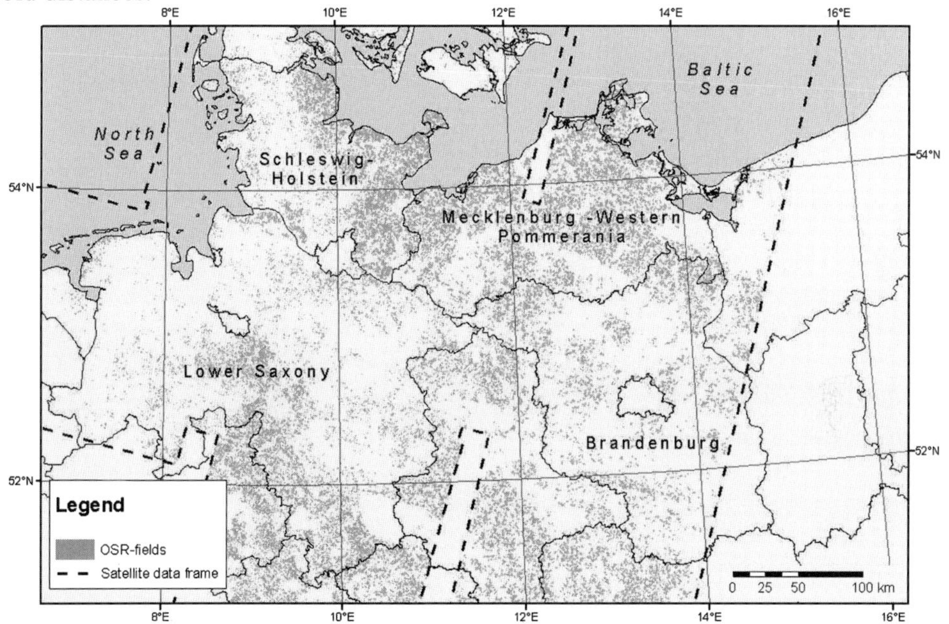

Figure 3.13: Automated detection of Oilseed rape cultivation areas in Northern Germany through a LANDSAT 7 ETM+ remote sensing image application in 2001 (data adapted from Laue, 2004).

Table 3.6: Number of field cases, average nearest centroid distance from neighbours, and field acreage within the respective Federal States of Northern Germany.

Federal States	**Schleswig-Holstein**	**Mecklenburg Western-Pomerania**	**Branden-burg**	**Lower Saxony**
Total area of federal state (Km^{-2})	15 763.4	23 178.8	29 478.1	47 619.6
Number of fields identified	9 516.0	8 985.0	4 922.0	15 202.0
Average centroid distances of nearest neighbours (m)	506.7	604.8	674.5	563.6
Standard deviation of centroid nearest distances	390.3	450.0	625.1	546.6
Median centroid distance of nearest neighbours (m)	411.9	490.4	496.2	416.9
Average field size (m^2)	63636.4	183225.2	145413.9	36513.4
Standard deviation of field size	109291.2	333759.3	248302.9	43938.4
Median acreage (m^2)	29700.0	45900.0	42300.0	22500.0

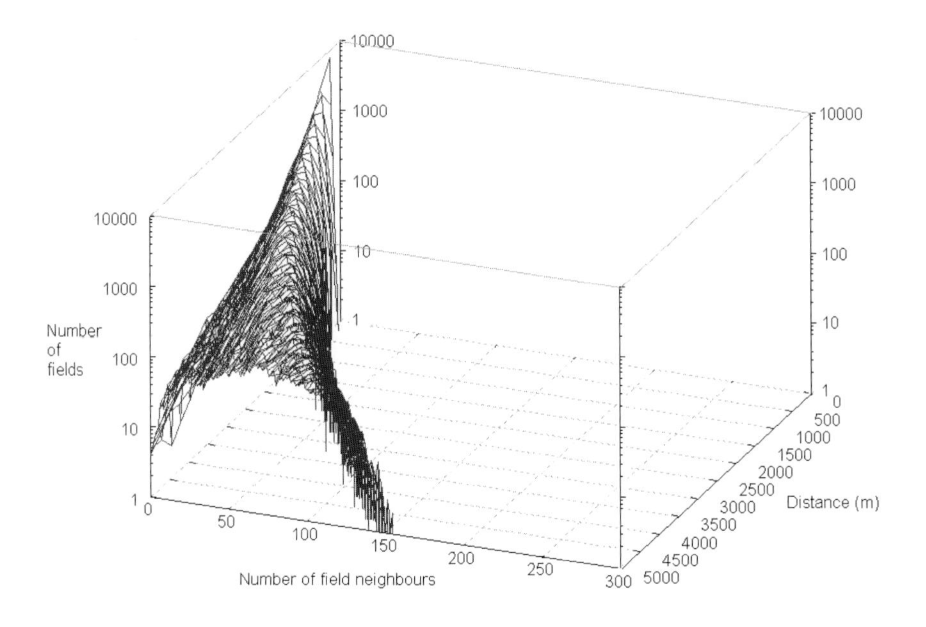

A) Schleswig-Holstein

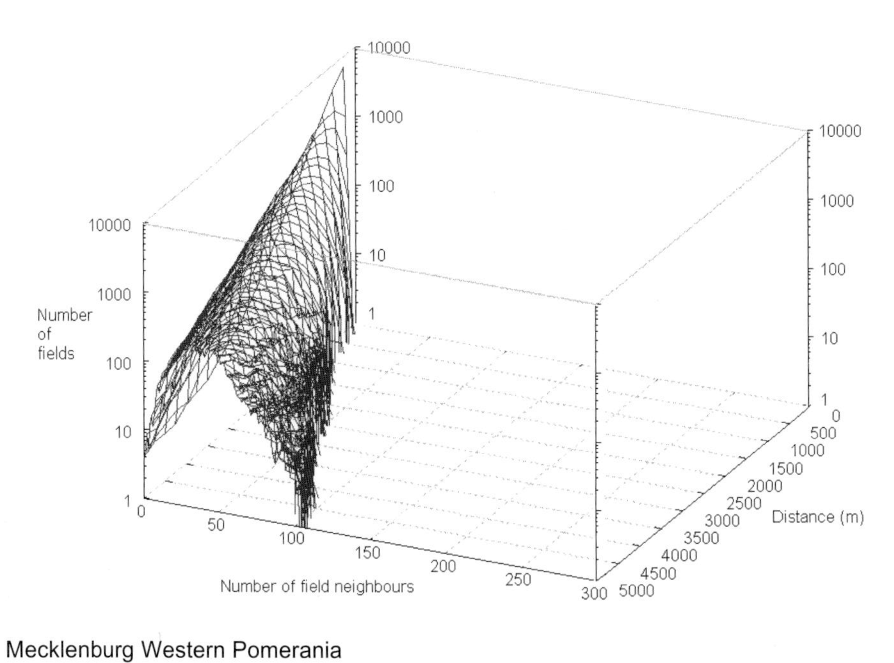

B) Mecklenburg Western Pomerania

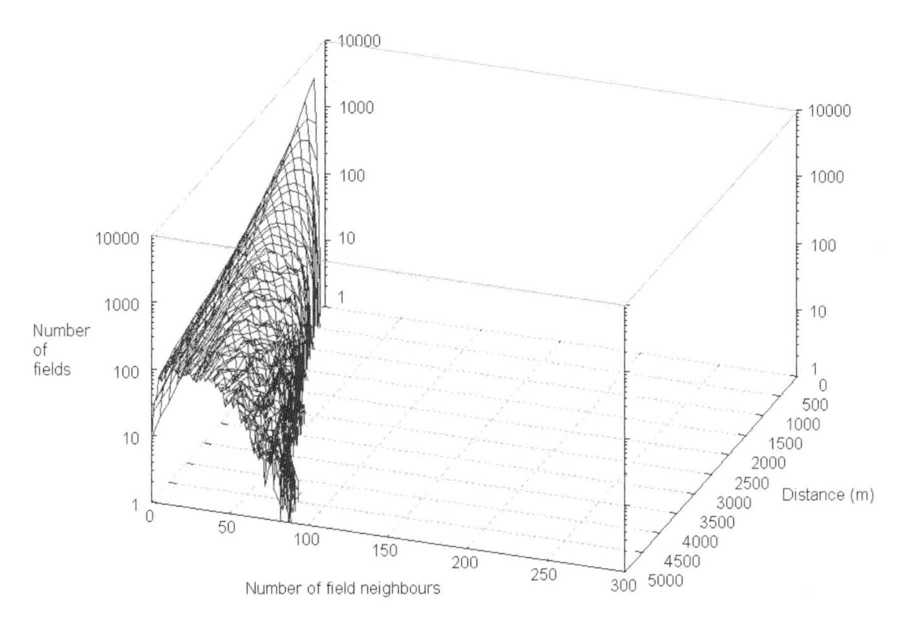

C) Brandenburg

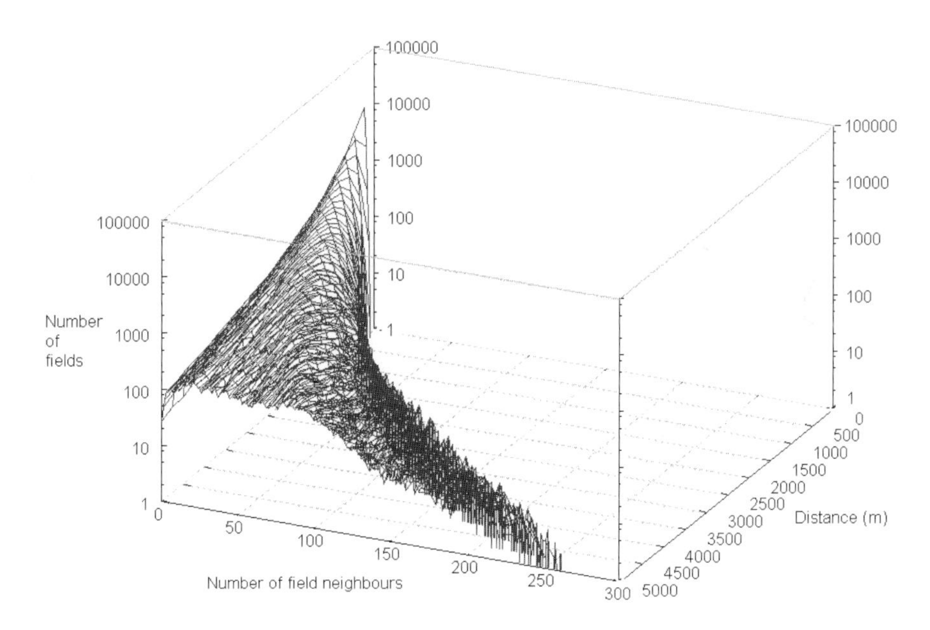

D) Lower Saxony

Figure 3.14 A-D: For the identified oilseed rape fields in the stated regions, it is shown how many fields were counted having how many neighbours at a particular distance (step-wise increased from 250 m to 5000 m in 250 m steps drawn on a logarithmic scale). For example Fig. 3.14 A shows that in shorter distances (e.g. within 500 m), there are only fewer neighbours less than 10 to recognize with respect to a specific field, and at larger distances, e.g. at 3500 m, 10 fields have about 100 neighbours on average. This also implies that more fields would be affected by isolation or monitoring measures.

3.2.4 Discussion

Pollen transfer between oilseed rape fields is possible across larger distances (Treu & Emberlin, 2000). For example, interaction processes involved in honey bee pollination activity range up to 5 km. Transgenic oilseed rape fields may contaminate adjacent conventional fields. This implies harvest impurities not only in transgenic but also in conventional fields. Seed losses during harvest may contribute also to a fraction of transgenic seeds in conventional fields with transgenic volunteer growth in subsequent crop rotations. The accumulation of transgenic genotypes with eventually stacked transformation events could complicate volunteer control and force multiple herbicide applications (Breckling & Menzel, 2004). The results presented here indicate, how many neighbours have to be taken into account depending on how large the interaction distances are assumed based on results from field-to-field interaction experiments. Regions with smaller fields and higher cropping densities have a higher biotic interaction and out-crossing probability with near neighbours as in the case of Schleswig-Holstein when compared to Mecklenburg Western-Pomerania for instance.

Devaux et al. in 2007 confirmed outcrossing in OSR via potentially long-distance pollen movements. Furthermore, smaller fields may have a higher probability of outcrossing with potential hybridization partners as they have longer field margins with wild natural habitats for the same cultivation area of larger fields (Laue, 2004). Taking into account also the potentially long viability of oilseed rape seeds in the soil seed bank (D'Hertefeldt et al., 2008); a potential accumulation of transgenic individuals over the years has to be considered. Farmers change boundaries of sub-plots cultivated homogeneously within the same fields, therefore long term estimates of whether farmers change their field locations during the years would require analysis of larger data sets (years/images) to get a broader picture.

3.2.5 Conclusions

Remote sensing application allowed to generate large-scale data on oilseed rape cultivation. Agricultural planning in the context of GMO cultivation requires the consideration of neighbouring fields. For coexistence planning as well as for estimations of relevant interaction sites for processes of a particular spatial extent it can be concluded how many sites have to be evaluated on average in a particular part of the considered region. This is an important issue to differentiate regional specificities. In particular, the empirical data are of practical relevance to estimate the efficiency of isolation distances and to address the implied coordination efforts in monitoring as well as in crop rotation planning among farmers.

This was the European part from which we have seen some examples, how to analyse the characteristics of cultivation systems on larger scales for an environmental assessment- now proceeding to look at African conditions.

Part III:
The Ghana case study on small-scale maize farming

Chapter 4: Background

4.1 The context of small-scale agriculture in Ghana and relevance for a biosafety monitoring concept in maize crop production

Ghana has a total land area of about 24 million hectares, of which agricultural land area constitutes about 14 million hectares (Ekboir and Dankyi, 2002). The agricultural sector including crops, livestock, fisheries and forestry accounts for about 40% of the country's Gross Domestic Product (GDP), employs in excess of 60% of the country's working population, and generates over 40% of the country's foreign exchange earnings annually. Agricultural food production is mainly subsistence in nature, with only a small proportion as large scale commercial enterprises. The subsistence nature of crop production implies that total crop output and marketable surpluses are subject to seasonal fluctuations as a result of rainfall variability and incidence of pests (Asuming-Brempong and Asafu-Adjei, 2000). Individual farm holdings measure between 1 and 2 ha (Al-Hassan and Jatoe, 2002), among which maize is the most widely cultivated cereal, as it constitutes a major staple for most communities (Fosu et al., 2004). In terms of seed exchange requirements, on-farm surveys have revealed that seed sources among small-scale farmers are to a large extent from previous harvests. Off-farm sources include procurement from other farmers as gifts. In addition, seed purchase is a relevant off-farm source (Walker et al., 1998). In relation to productivity of agricultural holdings, official statistics indicate that area annually planted with maize averages to about 650,000 ha (see Fig. 4.1d).

Over the years, maize acreage has expanded steadily into the overall agricultural land, estimating to nearly 6% of the total agricultural land area of the country (Fig. 4.1b). Most of the maize grown is cultivated in association with other crops, particularly in the coastal zones, so planting densities are generally low, thus affecting production outputs (Morris et al. 1999). Production outputs are largely constrained due to the high incidence of diseases, such as maize streak virus, infestation by the stem borer complex and the parasitic weed *Striga hermonthica*, as well as drought (Fakorede et al., 2003). Soil nutrient status is also very low (due to low soil nitrogen). The traditional slash and burn practices and the often indiscriminate bush burning further compounds the problem of soil fertility (Ekboir et al., 2002). Thus, attempts to increase agricultural productivity in the country has led to the observed situation, where the total area planted to maize per annum has increased at a faster rate (R^2= 0.8) (Fig. 4.1a) when compared to yield (R^2= 0.4) (Fig. 4.1c), in the longer term. Hence, average grain yields of maize are correspondingly lower when expressed per unit land area, averaging less than 1.5 t/ha per annum.

Maize production averagely estimates to 560,000 t per annum, and even though production generally increases, the upward trend is characterised by high inter-annual variability (Fig. 4.1d), most reminiscent of rain fed agriculture (Morris et al. 1999) and the incidence of pests (Fakorede, 2003). It is confirmed that such low annual harvests leads to a situation where maize production in the country falls short of domestic demand.

Thus the potential of different measures to increase agricultural efficiency in Ghana, and in particular for small-scale farmers is a key issue for socio-economic advancement and rural development. It is within this context, that the adoption of transgenic varieties, are discussed as a means to enhancing agricultural productivity in Ghana (Nuffield Council on Bioethics, 1999; ACDI/VOCA, 2003; BiotekAfrica News (2003); National Biosafety Framework for Ghana, 2004). However, the socio-traditional context of subsistence cultivation practices, seed exchange as well as the climatic conditions does not provide a direct estimation on how efficient a deployment of transgenic (Bt) maize varieties into small-scale farming systems would be.

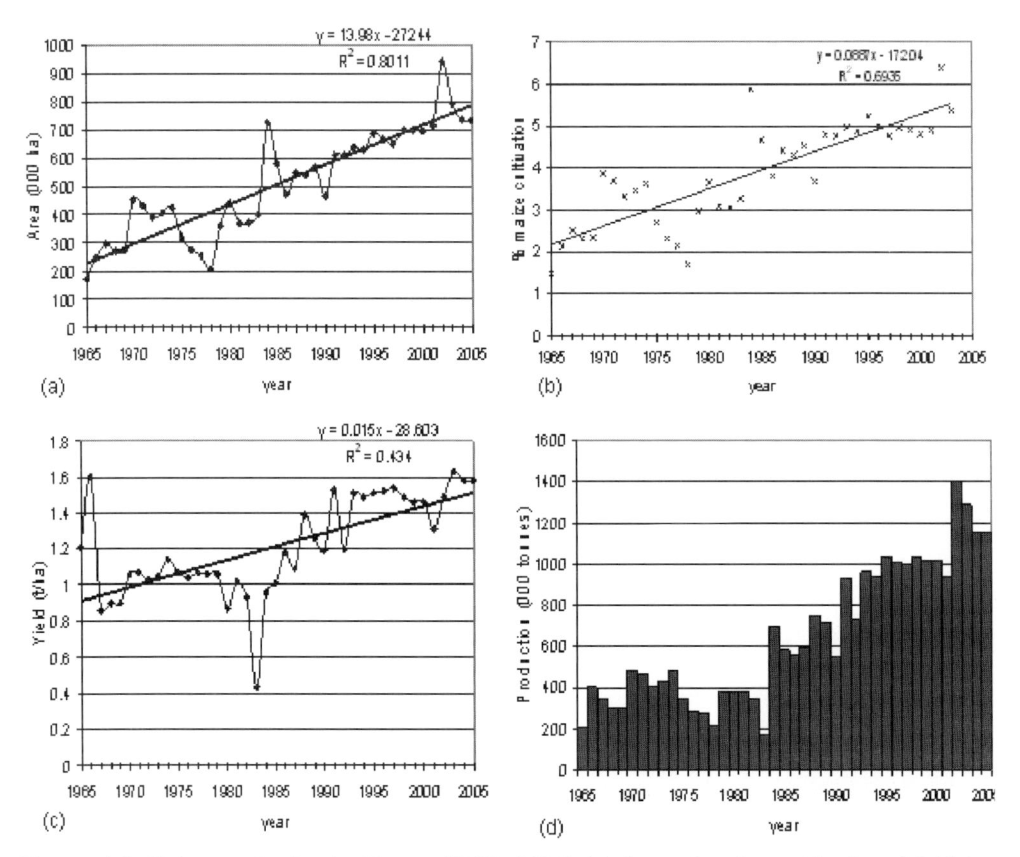

Figure 4.1: Maize production in Ghana (1965- 2005) (a) Area of maize cultivation; (b) Maize acreage as a percentage of total agricultural land area in the country; (c) Maize yield; (d) Maize production. Source: Calculated from FAO database (2007b); and Morris et al. (1999) Adoption and Impacts of Improved Maize Production Technology: A Case Study of the Ghana Development Project. Mexico, D.F.: CIMMYT.

Therefore scientific baseline information for an evaluation of the potential ecological risks and socio-economic implications of its adoption into smallholder agricultural practices is of high relevance for the biosafety development processes in the country. In this field remain important research deficits, which need to be filled as a basis for informed decisions. In this context it is necessary to anticipate effects of a potential deployment of transgenic varieties, based on the prevailing seed exchange and reproduction pattern of sample populations of conventional varieties, which are currently farmed. One alternative approach to understand maize pollen dispersal to predict risk of pollen moving off-site and cross fertilizing with ovules at a distance is the application of mathematical models, which is based on gene flow studies with respect to local environmental conditions (Brookes et al., 2004; Kuparinen, 2006; Eastham and Sweet, 2002; Aylor, 2004). Estimations on the agronomic and socio-economic conditions of traditional agriculture will provide information on relevant indicators for monitoring in relation to the prevailing seed reproduction pattern.

4.2 Specific terms of reference to transgenic maize contamination events in the past decade

The entry into force of the Catagena Protocol on Biosafety in 2000 provided a relevant framework for defining global biosafety regulation. Since concerns over genetic engineering were first raised in Protocol negotiations, there has been increasing evidence of ecological, economic and health risk impacts on farmers. In the following, global highlights of some major GM maize contamination incidences within the last 10 years are presented:

In 2008: Unapproved GM maize (MON 810) found in seed imports in Kenya: Maize seed imported from South Africa by Pioneer Hi-Bred was found to be contaminated with a GM variety – MON 810- that is not approved in Kenya. Greenpeace International, in cooperation with a coalition of several environmental and farmer's organisations in commissioned tests of 19 different seed varieties that were bought in seed stores from key maize producing areas across the country. The investigations which were conducted by an independent European laboratory, revealed that Pioneer's seed maize PHB 30V53, sold in the Eldoret region of Kenya was contaminated with MON 810 maize. GM maize MON 810 contains a novel gene that is considered unsafe and banned in several European countries (Greenpeace, 2008).

In 2007: GM maize contamination incidence in Spain: Fifteen organic farmers in Spain were affected by GM contamination through cross-pollination in 2003-2005. Contamination levels ranged from 0.03-12.6%, and in all cases organic certification was withdrawn, leading to financial loss as the grain could no longer fetch financial premium. One farmer lost €4000 due to the lower market price (Hewlett and Azeez, 2007).

In 2006: Farmers planting Bt maize in the Philippines: Filipino farmers in a province were misled into planting GM Bt maize. The province of Oriental Mindoro is an organic farming region that originally did not grow GM crops. Monsanto's local agent provided local farmers with generous loans to plant what was claimed to be conventional hybrid corn. Laboratory tests later confirmed the samples to be GM Bt maize (Greenpeace, 2008).

In 2005: GM Bt10 maize detected in imports in Japan: The first reported case of Bt10 contamination in Japan was in June 2005. During which a total of ten shipments of contaminated maize were detected. 32,610 tons of contaminated maize was placed in quarantine and returned to the USA. Syngenta had produced and distributed a variety of GM maize (Bt10) which did not have regulatory approval. Between 2001 and 2004, several hundred tonnes of the Bt10 maize were grown in the USA instead of a maize line Bt11. Subsequently, the breach was reported by the company to the US authorities, but was not made public until 3 months later. The problem was due to lapses in Syngenta's quality control procedures that did not differentiate between Bt10 and Bt11 with the resulting situation that Bt10 lines were used in breeding (Greenpeace, 2008).

In 2004: Maize seed contamination in Greece: Authorities in Greece discovered GM-contaminated maize on their market in 2004. The contaminated maize came in through seeds sold to farmers without their knowledge and had already been sown in various smaller regions in northern Greece. The Greek government traced and destroyed about 118 hectares of maize fields found to be contaminated with the GM maize. Pioneer and Syngenta offered various compensational sums to the affected farmers and made them to sign a contract. By that they were obliged not to oppose the company in anyway otherwise they would have to return the compensation amount they had received. Two separate law suits were pursued on behalf of small farmers organized by The General Confederation of the Greek Agrarian Association (GESASE) against Syngenta on the one hand and another against the Greek authorities, for allowing GM seeds to enter the domestic market (Greenpeace, 2008).

In 2003: Traditional maize contamination by GM varieties in Mexico: Twenty-five months after the first scientific evidence that Mexico's traditional maize is contaminated with DNA from GM maize, peasant farmers and indigenous communities in Mexico released the results of their own testing that found GM contamination of native maize in at least 9 states- far more widespread than previously assumed (ETC Group, 2003).

In 2002: US food aid contamination with Roundup Ready maize in Nicaragua: Sampling of US food aid found Roundup Ready maize, MON GA21, with contamination at levels of up to 2% (Friends of the Earth, 2008).

In 2001: Illegal planting of Monsanto GM maize discovered Argentina: In Argentina's three main maize production provinces of Buenos Aires, Santa Fe and Cordoba, huge amounts of GM herbicide tolerant Roundup Ready maize, GA 21 were planted. The main concerns were that they had not been approved for planting for human or animal consumption in Argentina, or for export to Europe (Greenpeace, 2008).

In 2000: StarLink Corn Products Liability Litigation in the United States: This represents the worst single contamination incident via StarLink maize, produced by Aventis (now Bayer). A group of corn farmers in the United States sued Aventis Crop-Science USA Holding, Inc. for alleged dissemination of a product that contaminated the corn supply, increasing farming costs and depressing corn prices, and they sought to recover damages on the basis of negligence, strict liability, nuisance, and conversion. Aventis had created a GMO variety of corn – known as StarLink – that was approved in 1998 by the Environmental Protection Agency (EPA) for certain use as animal feed but not approved for human consumption. The StarLink maize contains a gene from Bacillus thuringiensis, coding for an insecticidal Bt toxin known as Cry9C. This Bt toxin is not found in other GM insect resistant crops and there are concerns that it could be a human allergen because it is heat stable and does not breakdown rapidly in gastric acid in the human digestive system.

Main concerns are that Cry9C is not found in Bt preparations used directly as an insecticide, and that there is no experience with its use and safety. Aventis was therefore obliged that StarLink was segregated from other corn, that there was a buffer around StarLink corn crops to prevent cross-pollination with non-StarLink corn plants, and that farmers were made aware through a number of means of the conditions on the cultivation and sale of StarLink corn. In the fall of 2000, human food products were found to contain the Cry9C protein. As a result, many United States food producers stopped using United States corn and a number of countries terminated or limited their imports of United States corn. To date, the StarLink Bt variety approved only for animal feed has entered the human food chain in several countries including the US, Canada, Egypt, Bolivia, Nicaragua, Japan, South Korea, and in Mexico with episodes involving Mexican Taco Bell and corn sent to Central Africa as food aid, found to contain some StarLink corn with a consequence of a global product recall and billion dollar compensation paid by Aventis (CBD, 2006; Greenpeace, 2006; GMWatch, 2008).

In 1999: Maize seed contaminated with GM in Switzerland: Greenpeace (2008) reported that "The Swiss Department of Agriculture (Bundesamt fuer Landwirtschaft in Bern) discovered that Pioneer Hi-Bred's maize seed varieties were contaminated with Bt genes from a variety of maize genetically modified to be resistant to the corn borer. Prior to the discovery of the contamination incidence, Pioneer had sold seeds to sow 400 hectares (roughly 0.5% of total maize cultivation in Switzerland), about 200 hectares of which had already been planted".

In 1998: Cross-pollination by GM maize of neighbouring conventional crop in Germany: This was a situation where a neighbouring farmer did not know that GM maize was growing less that one metre from his field. The GM variety in question was produced by Novartis (now Syngenta) cross-pollinated the adjacent conventional maize field in the region of Baden- Württemberg, in southern Germany. Maize cobs up to 10 metres away from the GM-field were taken and samples analysed at the Freiburger Institut für Umweltchemie e.V. and GeneScan for Novartis. Analysis indicated that the rate of cross-pollination was around 5 % at the field border, 0,2 % at 5 metres and 0,1 % at 10 metres distance (Greenpeace, 2008).

Appendix 1.1 lists some transgenic food and feed products of maize as submitted for notification to the EU Commission in 2007. In the following, an overview of maize plant origin, production and utility is provided. Aspects of its reproductive biology and gene flow factors are further explained. In a final analysis, the context of urban agriculture is discussed with special reference made to the situation in Accra, Ghana.

4.3 The maize plant (*Zea mays L.)*

4.3.1 Origin

Maize is a member of the Gramineae (grass family). Originally from Mexico, it is now pan-tropical and also grown as a summer crop in temperate countries (ETC Group, 2003). The exact period of domestication and the ancestors from which maize arose are unclear. Archaeological records suggest that domestication of maize began at least 6000 years ago, occurring independently in regions of the southwestern United States, Mexico, and Central America (Mangelsdorf, 1974). In Ghana, maize has been cultivated since its introduction in West Africa in the late 16th century. It soon established itself as an important food crop in the southern part of the country. Earlier on, it never achieved the economic importance of traditional plantation crops, such as oil palm and cocoa. Over time, the eroding profitability of many plantation crops- attributable to increasing disease problems in cocoa, deforestation and natural resource degradation, and falling world commodity prices, served to strengthen the interest in commercial food crops, including maize (Morris et al., 1999).

4.3.2 Production

Maize is cultivated worldwide in diverse agro-ecological systems, with annual global production amounting to 709 million tons in 2005 (FAO, 2007b). Developing countries account for 64% of the world's maize cultivation area and 43% of the total global maize production. The United States is by far the largest producer of maize with an output of 282 million tons in 2005, followed by Asia, which recorded 195.5 million tons in the same year. Latin America and the Caribbean produced 86.7 million tons altogether, while in Europe an estimated 79 million tons was recorded in 2005 (FAO, 2007b). The FAO estimated least production figures in Sub-Saharan Africa amounting to 49.8 million tons in 2005, with South Africa representing the largest producer in the region at 11.7 million tons. Presently, maize is the most widely grown grain in Ghana (Walker et al., 1998), cultivated by the vast majority of households in most parts of the country (Morris et al., 1999).

4.3.3 Usage

Maize is grown for forage and silage, but mainly for its grain, which is borne in ears (or cobs), that arises from axils of the lower leaves. Maize represents a staple food as well as supplements the diets for a significant proportion of the world's population. Overall, 21% of the total world maize production is consumed as food, with consumption and utilization varying greatly around the world. In Mexico and Sub-Saharan Africa, maize is traditionally the main staple of the diet, i.e. 68% (Mexico) and 85% (Sub-Saharan

Africa) of the maize grown is used directly as human food. In general, maize grown in Southeast Asia makes up a small portion of the diet and is primarily produced for livestock feed purposes, except in Indonesia where maize is a major part of the diet (Morris et al., 1999). A nation-wide survey carried out in Ghana revealed that 94% of all households consumed maize during an arbitrarily selected two-week period. An analysis based on 1987 data for an example showed that maize and maize-based foods accounted for 10.8% of household food expenditures by the poor, and 10.3% of food expenditures by all income groups.

4.3.4 Reproductive biology

Maize is an anemophilous (i.e. successful dispersal depends on wind dispersal of pollen), annual species, monoecious i.e. spatially separate tassels (male flowers) and silks (female flowers) are found on the same plant (Fig. 4.2). In general they do not flower simultaneously, but the tassel starts the pollen release a day or two before the silk becomes receptive (known as proterandry). This feature makes cross-pollination common and limits inbreeding to a large extent. Overlap between pollen shedding and silk emergence can occur and up to 5% self pollination may occur in some cases. The ovary develops a long style or the silk, which extends from the cob and receives the pollen from the tassel. Maize cultivars and landraces are known to be diploid (2n= 20, 21, 22, 24), and are interfertile to a large degree. Pollination occurs with the transfer of pollen from the tassels to the silks. If fertilization occurs, individual kernels, or fruit formed are unique, in that mature seed is not covered by floral bracts, as in most other grasses, but rather the entire structure is enclosed and protected by large modified leaf bracts, collectively referred to as the ear (Hitchcock and Chase, 1951). The kernels are tightly held on the cobs and if ears fall to the ground, so many competing seedlings emerge that the likelihood that they will grow to maturity is low (Agbios, 2007). Maize cannot reproduce asexually by natural means. It is possible to reproduce maize using tissue culture techniques; however, it has proven extremely difficult with low success rates (Agbios, 2007).

(a) Pollen: Pollen is produced in the anthers in the tassel (male flowers) at the top of the plant and typically, has a diameter of 90 to 125 x 85 microns, a volume of about 700 x 10-9 cm^3 with a weight of about 247 x 10-9g (Della Porta et al., 2006). When mature, the anthers emerge from the tassel suspended on filaments and pollen is released from an opening at the tip of the anther. Maize pollen is released in very large quantities, between 4.5 and 25 million pollen grains per plant over a typical 5-8 day period (Brookes et al. 2004). Typically, major portions of the daily release usually occur during midmorning to midday (Aylor et al., 2003). They remain viable under natural conditions for about 24 hours, although this can fall to only a few hours in hot, dry weather conditions (Emberlin, 1999). Their relatively large size makes them heavy compared to other wind dispersed crop species, however, have a high terminal velocity resulting in higher comparative deposition (i.e. falls to the ground rapidly in a limited area and does not travel far in most of the cases) (Brookes et al., 2004).

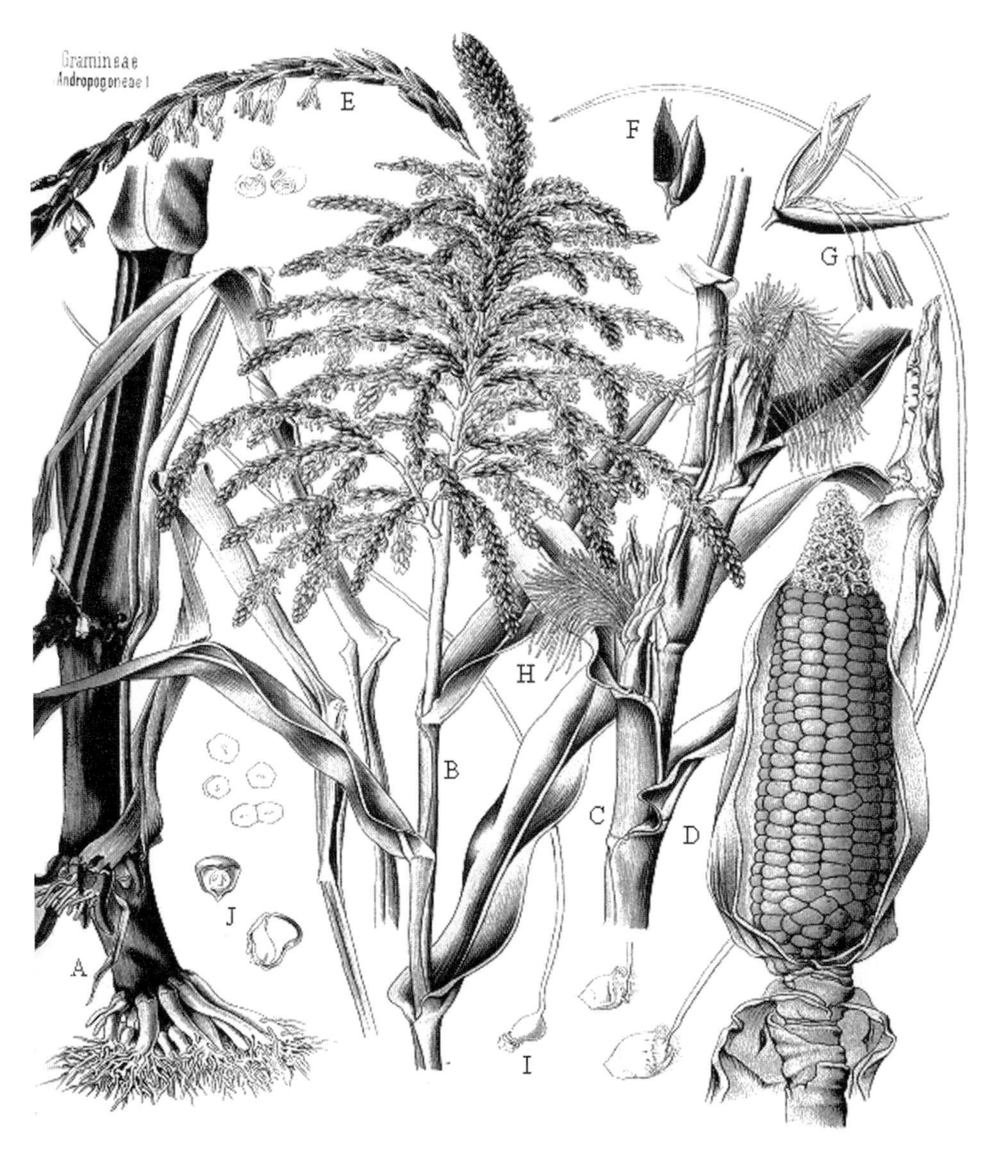

Figure 4.2: The vegetative and reproductive parts of the maize plant (Zea mays L.) A=stalk; B=leaf sheath; C=node; D=husk covering kernels; E=tassel; F=spikelet; G=male flowers; H=female flower (silks); I= anthers suspended on filaments; J= seeds (kernels). This figure has been reproduced with permission from the publisher (Kurt Stueber, Editor www.biolib.de).

(b) Silks: Silks are the styles of the female part of the flower. One silk directs the germ tube of a germinating pollen grain to one ovule, which must be fertilized in order for a kernel to develop. Silks emerge from the top of the ear and continue to grow until fertilized, some reaching lengths up to 15 cm or more. The silks themselves do not readily retain pollen, almost all of which is captured on trichomes (hairs) on the silks (Aylor, 2003). A single hybrid maize plant can produce within 6-8 days, 700- 900 silks. The ratio of pollen to silks is about ten thousand to one (Della Porta et al., 2006).

4.4 Studies on pollen mediated gene flow/ outcrossing

Gene flow is the 'incorporation of genes into the gene pool of one population from one or more populations' (Eastham and Sweet, 2002). Such gene movement is a major determinant of genetic structure in natural populations. Gene flow is strongly influenced by the biology of the species involved and is likely to vary with different breeding systems, life histories and modes of dispersal either through seed or pollen movement (Eastham and Sweet, 2002). However, the dispersal of pollen and seeds are key mechanisms leading to gene flow from GM populations (Kuparinen, 2006). It is under these circumstances that increasing cultivation of genetically modified crops has raised concerns over possible environmental and agronomic impacts if transgenes escape from GM populations and become established in natural or agricultural ecosystems (Messeguer, 2004; Eastham and Sweet, 2002; Snow et al., 2005). Potential environmental impacts may result through the movement of transgenes from crop to wild species or native populations. If crops hybridise with wild relatives and gene introgression occurs, wild populations could incorporate transgenes that could have adverse effects both for the conservation of genetic diversity and for plant breeding (Papa, 2005; Gray and Raybould, 1998).

Literature review into the dynamics of pollen flow and cross-pollination in maize[1] reveals that most maize pollen falls within 5 meters of the field edge. In the Sears and Stanley-Horn study of seven different Bt maize fields 84% to 92% of pollen fell within 5 meters and between 96% to 99% of pollen remained within a 25-50 meters radius of the maize fields. All pollen was deposited within 100 meters. Other studies have also analyzed the influence of size and shapes of fields, wind speeds and direction, as well as environmental conditions (Klein et al. 1998). Large rectangular fields result in pollen traveling further than small circular fields due to the higher concentration of pollen in the atmosphere at a given time. Also, the "depth of a field" and the direction of the wind is far more important than total area planted. As discussed, although pollen may be dispersed, it must be viable, land on the stigma of the receptor plant and compete with other viable pollen to be able to cross-pollinate (introgress):

- Salamov (1940) found cross-pollination levels of 3.3% at 10 meters from the pollen source, 0.5% at 200 meters, 0.8% at 600 meters and 0.2% at 800 meters;
- Jones and Brooks (1950) measured the percentage of outcrossing between large blocks of emitter and receptor crops over a three-year period. The average level of cross-pollination in rows immediately adjacent to the crop were found to be 25.4% falling to 1.6% at 200 metres and 0.2% at 500 meters;
- Jugenheimer (1976) found levels of cross-pollination of 4.5% at 3 meters;
- Burris (2003) found cross-pollination of 1.11% at 200 meters;
- Baltazar & Schoper (2002) identified no out-crossing beyond 200 meters in very dry and calm conditions;

1 Many of the references cited here (i.e. section 1.4.4) have been documented by Brookes et al. (2004) based on literature review done by Eastham K & Sweet J (2002) GMOs: the significance of gene flow through pollen transfer, European Environment Agency.

- In Bateman (1947), cross-pollination levels fell from 40% at 2.5 meters to approximately 1% at 20 meters;
- Messean (1999) measured 1% cross-pollination at a separation distance of 25-40 meters;
- Simpson (1999) found 1% cross-pollination at 18 meters from the pollen source;
- Loubet and Foueillassar (2003) showed that the fertilization capacity (% of pollen grains able to fertilize) decreases with distance from the source: 4%-12 % at 100 meters and 2%-7 % at 250 meters. This work also identified that the lightest pollen grains are the least viable, yet travel the longest distances and pollen placed in air flow (humidity 70 %) dies within 2 hours at a temperature of 20°C or within 1 hour at a temperature of 30°C.

Studies on Disruption of pollen dispersal and viability have shown the following: Jones and Brooks (1952) experimented with barriers to cross-pollination and found a single row of trees and under bush reduced out-crossing by 50% immediately behind the barrier. The reduction was even greater when an intervening crop was used (it provides competing pollen) and when open ground or low growing barrier crops exist to isolate maize crops, it appears that the first few rows intercept a high proportion of the pollen, so that cross-pollination levels are highest in these rows and then decrease with distance; Outcrossing rates tend to be higher at field edges than within a maize field of comparable size. Therefore, the use of mechanical barriers (like hedges, a line of trees) is only effective if established around a recipient field (Meir-Bethke and Schiemann 2003).

4.5 Factors influencing gene flow

Pollen dispersal and successful cross-pollination depends on several factors including pollen viability and its competitive ability, efficiency of pollination vectors, environmental factors, the size of pollen source and sink and the temporal congruence of pollen and receptor plant. The existence of buffer crops could also influence pollination processes. Another important factor that could lead to unintended presence through cross-pollination is the occurrence of volunteer crops. These factors are discussed as follows:

4.5.1 Pollen viability and competitive ability

Pollen viability (survival) is a key factor to effective pollen dispersal and fertilisation. Pollen survival is the ability of pollen to germinate, which stands in contrast to effective fertility, which refers to the ability of pollen to complete fertilization. This is because the former is a property of the pollen, while the latter is a property of the pollen plus the receptive state of the female silks and eggs. Hence the ability of pollen grain to function may be limited depending on where and when it lands. However, biological factors influencing successful pollination begin with the ability of the donor plant to produce viable pollen, and the length of time the pollen grains retain their potential for pollination.

If the competitive ability of the pollen grain is poor, its capacity to compete with fresher pollen produced in the vicinity of the receptor plant will be low. Pollen viability can vary greatly between species but is also dependent on environmental variables such as quantity of pollen, transmission mode, temperature and humidity factors (Eastham and Sweet, 2002).

4.5.2 Pollination vectors

Pollen produced by maize may be dispersed over long distances by both wind and insects. The weather can affect the behaviour of pollinating insects on the crop and the occurrence of air borne pollen movement so the amount of cross-pollination can vary significantly from crop to crop and the weather conditions of the day. The number and even species of natural pollinating insects can vary considerably in their contribution to successful pollination (Eastham and Sweet, 2002). Maize pollen transported on airflow over longer distances is likely to occur under a range of weather situations including uplift and horizontal movement in convection cells, and uplift and transport in frontal storms. As maize pollen grains remain viable for about 24 hours in normal weather conditions, pollination could occur at sites remote from the source (e.g. 180 km). Dispersal away from the vicinity of the crop also takes place by the carriage on bees. Hence evidence is cited that maize pollen is collected in notable amounts by bees. In this way, the pollen is transported several miles from the crop plot in suitable weather conditions to distant places (Emberlin, 1999).

4.5.3 Environmental factors

Pollen released on the airflow can settle by gravity, can be removed by precipitation, be absorbed into water droplets, or can impact onto surfaces including vegetation, buildings, soil and water bodies. The relative importance of these sinks and the impacts they have will vary with factors such as the terminal velocities of the pollen grains, climate, local vegetation and topography (Eastham and Sweet, 2002). Weather and changes in temperature, humidity, light, as well as wind and rain can heavily influence pollen dispersal. The drier and hotter conditions are at time of flowering, the lower the levels of cross-pollination and vice-versa. Levels of cross-pollination are highest in receptor crops that are typically downwind of donor crops. The stronger the wind at time of pollen dispersal, the greater the likelihood of cross-pollination recorded at greater distances (Brookes et al., 2004). Other factors in the local environment, such as nature of the plant canopy, surrounding vegetation and topography may influence the patterns of pollen dispersal. Wind velocity and airflow in particular, are affected by topography, and potentially affect movement from pollen source to receptor plants. In addition, physical barriers e.g. woods and hedges can obstruct air flow, having dual effects on depleting some pollen from the air by impaction, filtering and also creating a sheltered zone. Dense stands of shrubs, herb covers and tree-sized vegetation with full foliage acts as catchments for airborne particles, including pollen (Eastham and Sweet, 2002; Treu and Emberlin, 2000). In some cases, obstructions could divert pollen upwards and hence could travel further than otherwise would be the case (Brookes et al., 2004).

4.5.4 Buffer crops

The planting of (non GM) buffer crops affect cross-pollination levels. This is because a non GM buffer crop (of maize) can act as an interceptor to a large proportion of GM pollen and can provide additional non GM pollen that 'crowds out' the GM pollen (further reducing the chances of the GM pollen introgressing with the non GM crop in which adventitious presence is to be minimized).

4.5.5 Size of pollen source and sink

The cross-pollination rate from one field to another has been shown to depend on the sizes of both fields. If pollen disperses from a small source area it may behave as a narrow and unpredictable diffusion cloud. Evidence has shown that most airborne pollen from small to moderate sized fields contributes to the local component in this way. For instance studies have confirmed that a square 400 m^2 crop would emit ¾ the amount of pollen that a 4 ha (40 000 m^2) crop would emit, but suggested that the effectiveness of pollen dispersal would decline significantly in crop areas less than 400 m^2 (Eastham and Sweet, 2002). However, it is important to note that there could be enormous variability in all parts of the process.

4.5.6 Temporal congruence of pollen and receptor plants

Whenever there is a temporal alignment of pollen shed and silk emergence in a field, pollen grains from local sources greatly outnumber pollen from adventitious sources. On the other hand, when such temporal overlap becomes limiting, there is increased opportunity of pollen from outside sources to fertilize the ear. Adverse environmental conditions, such as drought, can alter the timing of silk emergence in relation to pollen shed (Aylor et al. 2003b). Therefore, it is important that some overlap in flowering times between the pollen donor and the receptor plant occur, so that ripe pollen and receptive stigmas are produced at the same time. In this case, a higher degree of cross-pollination might occur then if partial self-pollination had begun in one of the plants (Eastham and Sweet, 2002). Planting times and the time each variety takes to flower (and produce/ be receptive to pollen) usually vary by variety. Consequently, varietal differences can contribute differently to the timing of flowering and hence to the chances of cross-pollination occurring.

4.5.7 Volunteers

Maize is a domesticated plant not capable of disseminating its seeds by itself but relies on man as the main agent for its dispersal and distribution. However, the presence of volunteer plants from an earlier crop may increase the level of unintended presence in a crop. Whilst this possible source of unintended presence is potentially highest in regions which do not have low average winter temperatures to kill volunteer plants. Farm level experiences, e.g. in Spain, shows that this is an existing source of adventitious presence in maize (Brookes et al., 2004).

4.6 Hybridization and gene introgression

Hybridization is the cross-breeding of genetically dissimilar individuals. Such individuals may differ by one or a few genes (the pure lines of plant geneticists), by several genes (e.g. hybrid maize) or be very different genetically (as in most hybridization between members of different genera (Eastham and Sweet, 2002). Within species, hybridization is common but can also occur between species and occasionally with species in different genera. Hybridization between species is described as "interspecific" hybridization or where species belong to a different genus "intergeneric" hybrid. However, the incidence of natural interspecific and intergeneric hybrid varies substantially among plant genera and families (Eastham and Sweet, 2002). Plant interspecific hybridization is a common means of extending the range of genetic variation beyond that shown by the parental species. However, inherent problems of interspecific introgression such as hybrid instability, infertility, non-Mendelian segregations, and low levels of intergenomic crossing-over can constitute important limitations. Moreover, polyploidy or ploidy of dissimilarity between species may result in supplementary facilitation for interspecific gene flow (Herrera et al., 2002).

Hybridization is a frequent and important component of plant evolution and speciation, although the resulting F1 (first generation) plants are often sterile and relatively few populations persist, except where the parents remain in contact or where they are able to spread vegetatively. However, the frequent occurrence of fertile hybrids increases the chances of introgression, the incorporation of alleles from one taxon to another, mediated through repeated backcrossing of hybrid individuals to one of the parents (Eastham and Sweet, 2002). Without considering introgression due to modern plant breeding, gene flow from wild to domesticated populations may also occur but only when farmers use part of their crop production as seed for the next generation of planting, without replacing the seed with that of commercial varieties. This situation generally occurs in traditional agriculture by use of landrace populations. However, in the context of modern agriculture, gene flow may act in only one direction, that is from domesticated to wild populations (Papa, 2005).

4.7 Urban agriculture in the developing world

Urban agriculture (UA) is the growing of plants and the raising of animals for food and other uses e.g. small-scale gardening and forestry activities; and related processing and marketing activities, within and around cities and towns. It plays an important role in urban poverty alleviation, enhance social inclusion, improve upon urban food security, help in urban waste management as well as facilitate urban greening (Van Veenhuizen, 2006). The experience of urban farming in countries across Africa, Asia and Latin America underscores its increasing importance as an economic activity, with major significance for food security, stable family incomes and liveable urban environment. Thus the practise has been embraced and promoted by the international development community as a means for urban dwellers to achieve sustainable local livelihoods (UNDP,

1996). The risks of harassment and crop destruction by local authorities, loss through theft and predation, and other drawbacks are outweighed by the perceived advantages and gains from urban cultivation (Obosu-Mensah, 2007). It is reported therefore that many low-income households who farm in cities in the developing countries gain a more consistent source of food and earn or free up cash for non-food items (Hovorka and Lee-Smith, 2006).

It has also been confirmed that food production in urban and peri-urban areas creates stronger local economies by creating jobs (see Wikipedia, 2007). It is acknowledged therefore, that the practice does not only support a more sustainable production of food that tries to decrease the use of harmful pesticides (Wikipedia, 2007), but also serves as a model for the inclusion of different urban sub-communities into a social unit (community capital), for the realisation of several other community objectives (Smit and Bailkey, 2006). In this regard, it is observed that many governments are creating agencies to help support the practice related to the provision of land and other production inputs, as well as technical assistance. For instance, in greater Bangkok, 60% of the land is under cultivation, 72% of all urban families are engaged in raising food, mostly part-time (see City farms, 2007). Argentina, Peru and Mexico have metropolitan agencies that promote urban agriculture. In Africa, Côte d'Ivoire, Malawi and Tanzania are examples of countries that have set aside areas for urban agriculture, while in Nigeria and Zaire, urban farmers have been protected and encouraged through land use regulations and tax concessions (Dima et al., 2002). It is the aim of this study to analyse implications of GM maize within this sector of agriculture, which has been left out of consideration in the usual biosafety studies and is one of the important typical situations being relevant for many developing countries but not existent in a comparable form in those countries, for which GMO have been subjected to the most extensive biosafety studies.

In the following chapter, the methods used in the study are provided. These cover aspects of spatial analysis involving the utility of the GPS and GIS systems to study the local cropping system, crop demography, approaches used in modeling maize cross-pollination and the socioeconomic assessment.

Chapter 5: Methodology

5.1 The study area of Accra

The study was carried out in 2006 in Accra West, a peri-urban area of the Greater Accra Region (Fig. 5.1), where in recent years small-scale agriculture has been observed to play an increasingly important role in urban food production largely among the informal sector population (see Obosu-Mensah, 2007). Accra is the capital city of Ghana, the smallest and most densely populated of the 10 administrative regions in the country. It has a population size of about 2,905,726 (National census, 2000) accounting for 15.45% of the total population. The city lies in the coastal savannah zone with total land surface area of 3,245 km^2 or 1.4% of the total land area of the country (Ghana Districts, 2008). The population growth rate is estimated at 3.4% per annum in the city itself but up to 10% in its peri-urban districts (RUAF, 2007).

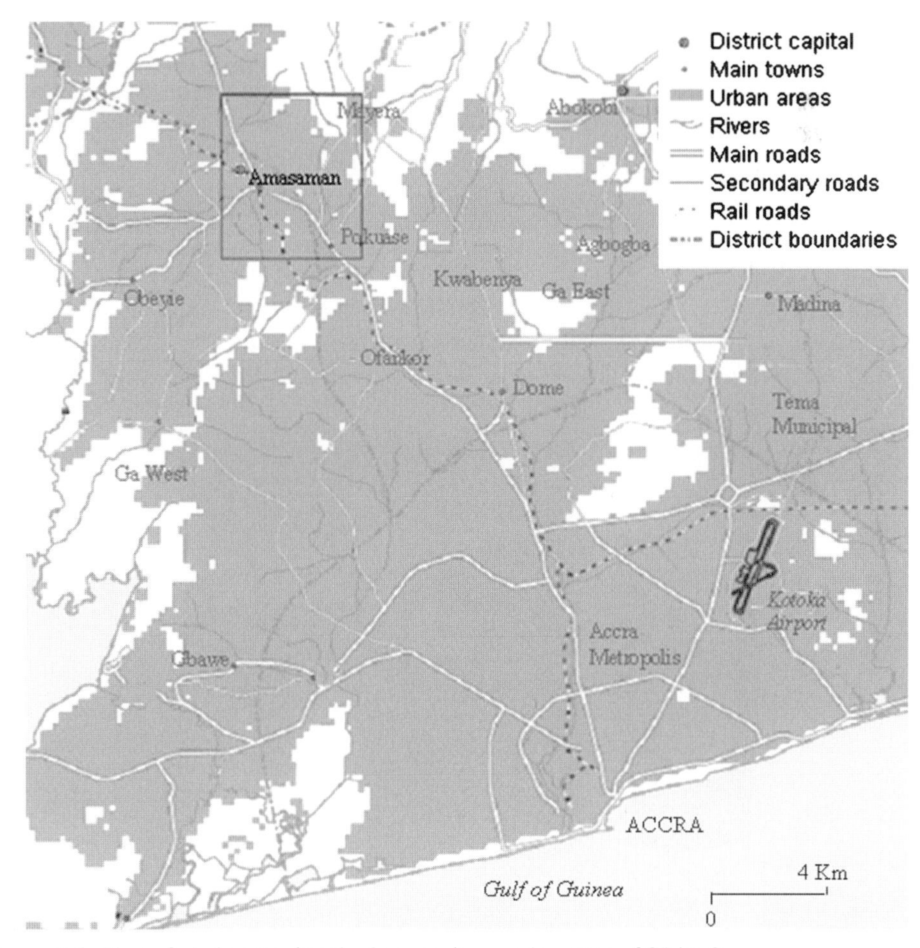

Figure 5.1: Map of study area (marked square), covering area of 25 km2.
Source: World Bank, 2007
http://siteresources.worldbank.org/EXTINSPECTIONPANEL/Resources/GHA35710.pdf

Two major categories of urban agriculture in Accra have been identified namely: household gardening, which takes place in and around homes and open-space farming, largely for semi-commercial reasons, taking place on lands some distance away from home dwellings. Home gardens usually involve small-scale farming on own premises or rented compounds and carried out typically for home consumption. However, other miscellaneous minor farming systems do exist e.g. micro-livestock and snail farming, bee-keeping as well as rearing of large ruminants (Bennet-Lartey et al., 2007).

In the next section, a description of the specific methods is given. Firstly, are the procedures used in the ground surface data gathering, maize mapping procedures and type of spatial analyses employed. The protocols used for capturing data on cropping frequency and phenological classification are briefly described. The Maize Model is then presented and followed by a description of how the socioeconomic conditions involving smallholder farmers were surveyed.

5.2 Spatial analysis with the help of Geographic Information System (GIS)

5.2.1 Concept of the Global Positioning System (GPS)

A GPS (Global Positioning System) receiver was the main tool used to gather spatial data on maize cultivation in Accra (Fig. 5.2 below). The GPS is a satellite-based navigation system made up of a network of 24 satellites orbiting the earth as high as 12,000 miles installed by the U.S. Department of Defence. The satellites move constantly with two complete orbiting turns in less than 24 hours (speed of about 7,000 miles per hour), and powered by solar energy with additional backup energy provided via batteries.

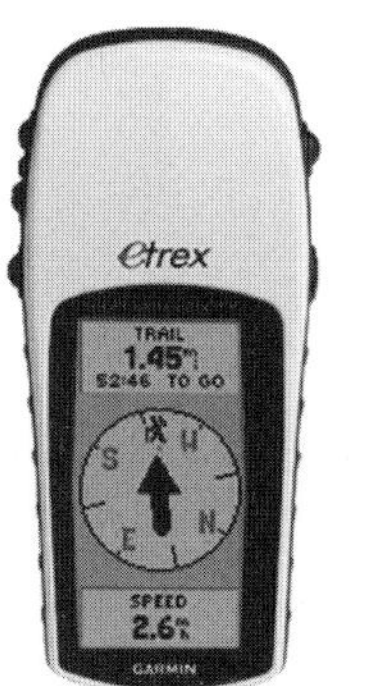

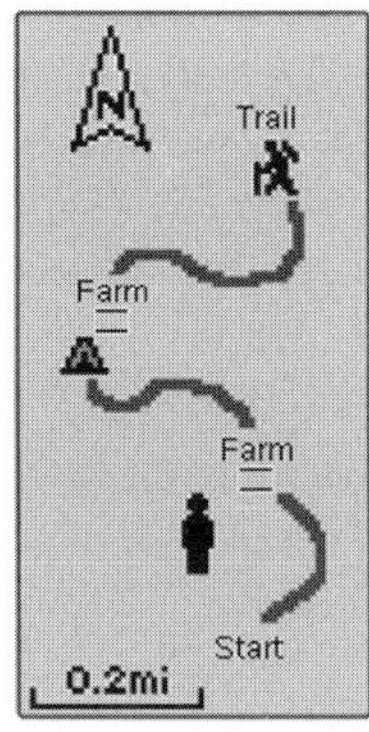

Figure 5.2: GARMIN ETREX H Handheld GPS Receiver used in the surveys and screenshots of pass points. The instrument is a high-sensitivity receiver, maintaining its GPS location even under heavy cloud cover or deep canyons, deep tree vegetation or near tall buildings. The TrackBack feature enables to reverse track logs to navigate back to a starting point (Source: http://www.garmin.com/garmin/cms/site/us).

Small rocket boosters on each satellite keep them flying in the correct path. GPS receivers access information from a minimum of three satellites to provide an accurate 2D location (latitude and longitude) or pass points of the user. A user's 3D location (latitude, longitude and altitude) is determined when the receiver obtains signals from at least four satellites in view. Additionally, it could provide data on date, local time, and could store up to 500 waypoints in memory for subsequent retrieval. The receiver triangulates the signals to estimate user's precise position by calculating the time difference between signal transmissions by satellite with the time it was received. With distance measurements from additional satellites, the receiver determines the user's position and displays it on the unit's digitized map (See: Garmin, http://www8.garmin.com/aboutGPS/).

5.2.2 Ground surface data gathering

Mapping of farms and feral locations was undertaken based on 1 km^2 grid maps specifically developed for this purpose (Fig. 5.3) using the GARMIN ETREX GPS instrument. The operational parameters were set for the local conditions and map transects were followed in cyclical successions per km^2. Measurements based on daily schedule of activities covering all homes and publicly accessible areas (e.g. Fig. 5.4). Three investigators were involved in data collection from May-September 2006. Investigations were done with the help of a car, extensive use of bicycles, and use of foot where possible along the demarcated routes by:

- Marking the precise locations of maize stands
- Documenting the list of waypoints
- Managing the routes

Table 5.1: Ground surface data based on GPS measurements.

S/n	Assignments	Descriptors	Relevance
1	Field, plot or garden locations	Specification of single locations of allotments based on GPS readings of first point of entry of the cultivation area	For estimating minimum distance between fields. This is an important parameter for estimating the probability of gene transfer (pollen) from modified to conventional maize fields. This is also useful to estimate the length of field borders.
2	Field, fields or garden sizes	Estimation of total acreage of fields/ plots- measurements taken at corners of the cultivation area ranging from 3-22 corners, depending on fields extent	Mean field size gives information on the dispersal characteristics of the cultivation area. The spreading of pollen is likely in regions with large number of smaller fields than in regions with fewer larger fields.
3	Feral/ volunteer locations	Specification of precise location points within same habitat patch	For estimation of nearest neighbour relations. Assesses the probability of cross-pollination between fields and feral locations

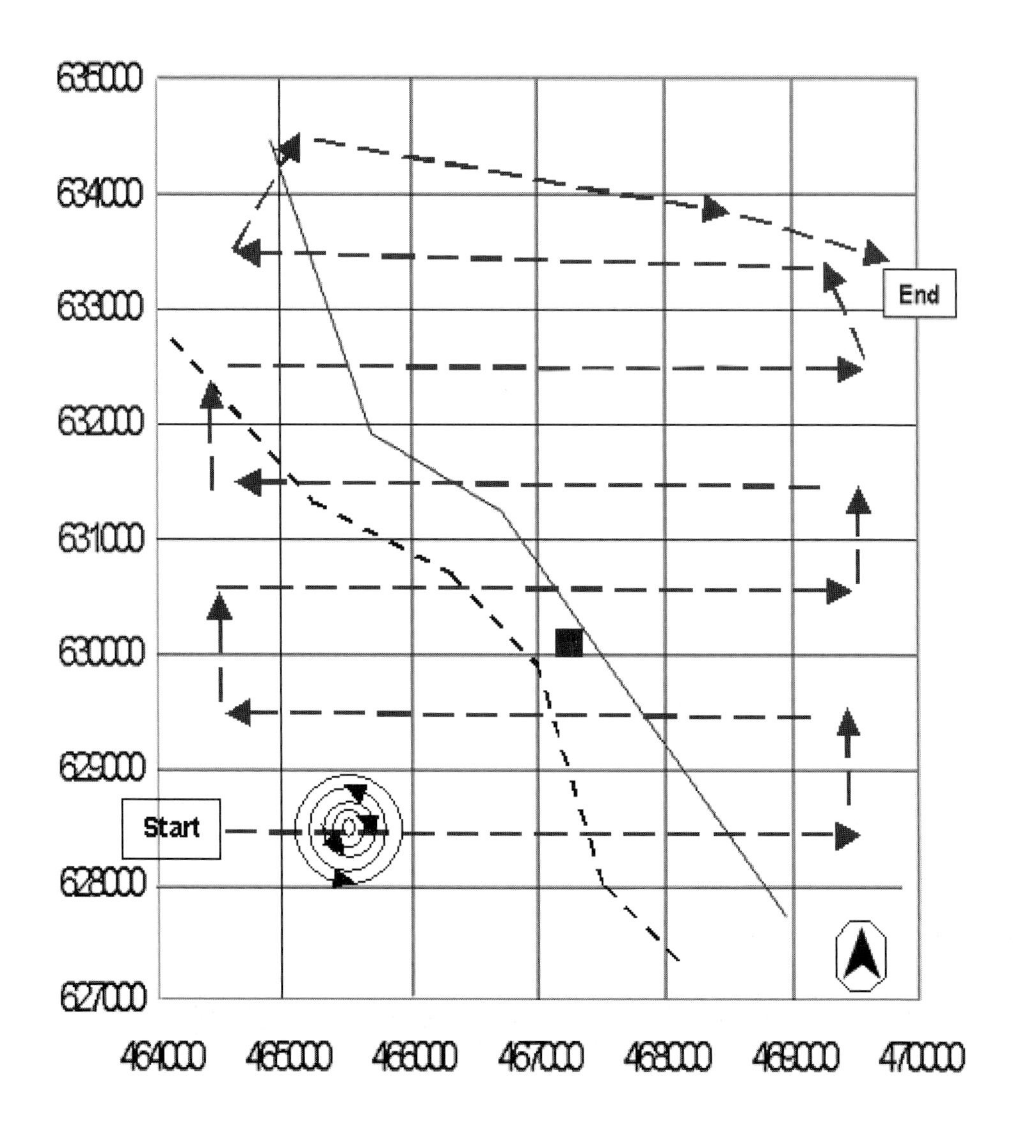

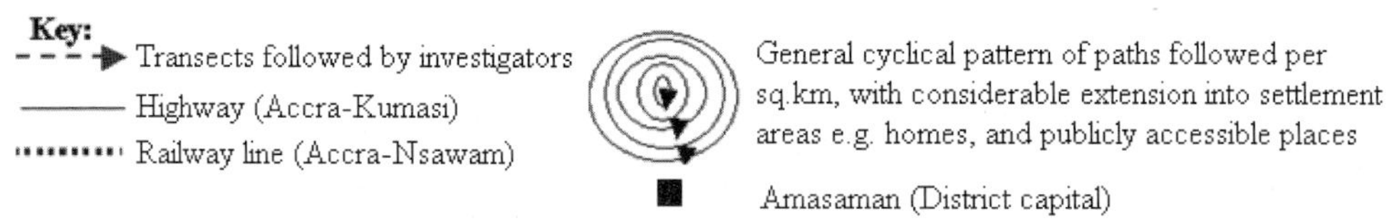

Figure 5.3: Transects of mapping routes followed by investigators within 1 x 1 km grids in Accra West (2006) based on a Gauss-Krüger coordinates. To ensure a complete coverage of study areas within the grid maps, cycling was done through all publicly accessible areas, homes, open areas, along roads and highways among others, including foot visits e.g. into wetland places and sand winning sites. Each grid was exhausted before proceeding to the next grid, until all 25 grids representing 25 km^2 were mapped. The methodology followed Menzel, 2006, and SIG-MEA Project (2005).

The location data were characterized basing on the following 3 assignments as shown in Table 5.1.

The acquired data were further mapped on Google imagery with a resolution of 1 x 1 km (Google Maps, 2008). This was the most feasible option available at the evaluation stage since satellite remote sensing data for the Area of Interest (AOI) to produce maps of identified crops by pixels is difficult to establish. Overlapping data frames from QuickBird satellite imagery of the area of interest (AOI) do not exist in the image archives for the past decades. Specific order attempted for this study by Digital Globe [via Eurimage, Italy, www.eurimage.com] in a window period from November 2007- March 2008 was not without cloud-free cover, hence the application of Google imagery obtained for the AOI.

Figure 5.4: Mapping of some cultivated locations in peri-urban suburbs of Accra (a) Maize farm at Amasaman; (b) Backyard garden at Amasaman; (c) Garden in an uncompleted building at Fise; (d) Home garden at the Legion village.

Table 5.2: Mapping of farm locations and corners (outlines) of cultivated areas.

Variables	Description
(1) Name of investigator	Name of person (s) who recorded the data
(2) Date	Date (dd/mm/yr) on which the data was recorded
(3) Name of community	Sub-urban community of farm location
(4) Data sheet No.	The numerical assignment of data sheet
(5) Field No.	Serial code for that particular field
(6) Field location	GPS coordinates of first point of entry to field
(7) Field corners	GPS coordinates field corners

Table 5.3: Descriptors for plant locations.

S/n	Finding locations	Site descriptors
1	Home compounds	Front and backyard gardens, farms in home neighborhood, around hotels
2	Traffic areas	Along railway tracks, highways, footpaths, road verges
3	Wetlands	Farms along water ways, fish ponds, streams, ditches
4	Marginal sites	On mounds, sand heaping, waste disposal sites
5	Construction sites	Farms in uncompleted buildings, house foundations,
6	Privately-owned open spaces	Farms on uninhabited private lands, fenced areas
7	Public areas	Farms on government lands, churches, mosques, schools, clinics, government offices, sporting areas
8	Small business centers	Farms around fitting shops, provision kiosks, milling places, animal husbandry,
9	Industrial/ factory sites	Farms around saw mills, fuel filling stations, brick-making, etc.

The mapping sheets are placed in Appendices 2.1.1 and 2.1.2.

5.3 Crop fields, volunteers, feral/ wild relative demography

5.3.1 Cultivation density

The cultivation density provides a direct measure for the number of plants present in a region. This parameter is linked to the pollen density or where relevant to the seed dispersal (Laue, 2004). Hence an estimation of the number of individual plants within a specified area of cultivation, or habitat patch was done following Menzel, 2004 (Table 5.4).

5.3.2 Classification of varieties grown

Name of varieties cultivated was obtained from farmers, where possible. For coding purposes, a classification of commercial varieties based on list of maize varieties and hybrids developed under the Ghana Grains Development Project (GGDP) following Morris (1999). See Appendix 1.2.

5.3.3 Phenological classification

This relied on maize growth stage classifications, from the seedling (cotyledon stage) to maturity levels, elaborated from CIMMYT maize growth developmental stage classification (2007). Table 5.5 provides basic protocol for population demography estimations for the region. Fig. 5.5 presents a pictorial overview of the classification.

5.3.4 Vitality of maize stands

This was used as a measure to estimate the 'well-being' or 'impaired' states of the populations, classified as being 'poorly developed', 'developed' or 'weakened by visible damage' etc., following a classification developed by Menzel (2006) applied in the context of oilseed rape (*Brassica napus*) and related species. In Table 5.6, the parameter descriptions are provided.

Table 5.4: Parameter description for frequency, degree of covered surface & size estimations.

Code	Number of individuals	Cover values (weeds)	Size of site m^2
1	< 100	Open surface, only single plants (below 10 % coverage)	≥ 2000
2	100-500	Plant coverage with gaps (10 to 40 % coverage)	2000-4000
3	500-1000	Open surface and coverage have similar values (40 to 60 % coverage)	4000-6000
4	1000 -1500	Small gaps (60 to 90 % coverage)	6000-8000
5	1500-2000	Surface nearly completely covered (90 to 99% coverage)	8000-10000
6	> 2000	completely covered, no bare ground to be seen (100 % coverage)	>10000

Table 5.5: Phenological descriptors of maize stands.

Code	State	Code	State
(1)	Seedling (only embryonic leaves/ cotyledons)	(5)	Mature, unripe (green) fruits/ cobs
(2)	Beginning of longitudinal growth	(6)	Ripe (yellowing) fruit/ cobs; Completely withered silks
(3)	First appearance of tassels without fruit sets	(7)	Dry fruit sets/ brown husks
(4)	First fruit sets (immature)	(8)	Open husks and scattering seeds

Table 5.6: Parameter descriptors for the vitality states of plants.

Code	Vitality
0	Poorly developed
1	Developed
2	Well developed
3	Extraordinary developed
4*	Weakened by visible damage: * Visible damage A) by pest: i. Feeding traces e.g. birds, not determinable; ii. feeding traces: stem borers iii. Feeding traces: army worms; iv. feeding traces: stalk borers B) by disease: i. Corn smut; ii. Rust; iii. White leaf blight; iv. Maize steak v. Mildew; vi. not determinable; C) Other: i. knocked down (by vehicles); ii. knocked down

Figure 5.5: Commercially-oriented fields within available open spaces (a and b) and Subsistence-oriented farms in spatially restricted areas within living areas (c-d).

5.3.5 Farm type

Estimations of the farm type were made based on relative size dimensions presented in table 5.7.

5.3.6 Characterization of feral stands

(a) Identification and mapping: The feral stands found were mapped based on the same sampling scheme for cultivated areas (Table 5.2 & 5.3). Feral stands found (e.g. see Fig. 5.6) were mapped using similar mapping procedure for the fields.

(b) Estimating fate of feral stands: This constituted a second phase investigation whereby volunteer and feral locations found were revisited in a second instance within a space of 2-3 month period, to ascertain their viability, presence or absence. In cases where they were still present, crop demographic data were again recorded.

Figure 5.6: Feral maize stands growing on the side of highways, roads and other crop fields. These plants presumably emerged as a result of seed fall outs from transport vehicles; (a) within road gutters (circled); (b) within a beans farm; (c) occurring in an open settlement area; (d) along a highway.

Table 5.7: Parameter descriptors for the farm types found.

Farm type/ orientation	Estimated size range dimension (Ha)	Code
Subsistence plot (Gardens and fields for family-self supply)	<1-2	50
Commercial plot (Gardens and fields for market supply)	> 2.0	60

5.3.7 Photo documentation

In principle, two pictures were taken at each location, namely the plant and another of the plant together with its site. These were aimed at supporting a second phase mapping exercises and to serve as a template to re-check aspects of the population dynamic data.

Table 5.8: Photo documentation of plant stands location.

Photo designation e.g.	Description of code
Ms1.1	First picture taken of a maize stand (farm) with a background view of the site of occurrence.
Ms1.2	Second picture taken at the same site (i.e. s1)
Ms1.p1	First picture of an individual plant at the site
Ms1.p2	Second picture of the same plant (i.e. p1)
Others	Other observations of interest e.g. flowering situation, pests, diseases, etc.

5.4 Modeling methods

5.4.1 The Maize Model (MaMo)

To analyze cross-pollination at the regional level, the Maize Model (MaMo) was applied (Fig. 5.9). The model was developed at the UFT by the Joint Working Group on the GeneRisk Project. The approach predicts on a regional scale, cross-pollination rates between GM and conventional maize fields for specific situations. Among others, it bases on available studies on maize outcrossing probability with distance, field sizes, and field numbers selected according to relevant scientific criteria (Reuter et al., 2008). The model concentrates on studies with modern hybrids which have relatively lower outcrossing rates. This is because modern plant breeding has reduced proterandrie, i.e. the time difference between pollen maturation and the silk (female) maturation of the plants. Only studies with source and receptor field of comparable size (difference no more than 1:15) were used (Reuter et al., 2008).

The resulting compilation of measured values is shown in Figs. 5.10a & b. The empirical data allowed to derive a function describing the mean quantity of cross-fertilisation related to distances (median). This value was calculated within a moving window. The double logarithmic display provides a better overview since from a few meters to a few kilometers gene flow differs in several orders of magnitude. The function together with the derived variability was employed in Monte Carlo simulation experiments by taking each field as a source and as a receptor in cross-pollination. Repeated model runs distribute pollen between all fields, deriving quantitative estimations of pollination rates and based on the synchrony effects of the flowering phase. The model is useful to help analyse GM-Maize cultivation scenarios on a regional level, and to further support mitigation measures and analyse socio-economic implications (Reuter et al. 2008). Quantitative models have been applied because they are particularly useful for systemic risk assessments and when a gene flow process should be investigated over a large variety of environmental and farming conditions (Messeguer, 2004; Boch et al., 2002, Kuparinen, 2006).

5.4.2 Model parametisation

The model was largely developed for the European conditions. Therefore, it was necessary to derive relevant parameters of the local Ghanaian conditions for the model adaptation. Thus, the following vegetative growth period was specified:

- the duration of pollen overlap between conventional and simulated GM varieties;
- the maximum distance of pollen dispersal; and
- the duration of vegetation phase prior to flowering as emprically determined under local conditions are provided in Table 5.9 below.

Table 5.9: Local variables used in the model scenarios.

Parameters		Mean duration of pollen shed (days)	Minimum duration of pollen shed (days)	Maximum duration of pollen shed (days)	Standard Deviation
Trait	Conventional and landraces	9.53	8.58	11.00	0.77
	GM and Commercial hybrid varieties	9.15	8.42	10.50	0.64
Maximum distance for dispersal of pollen (m)		4,500			
Mean vegetation period until flowering (days)		71.75			
Minimum vegetation period until flowering (days)		42.00			
Maximum vegetation period until flowering (days)		119.00			
Standard Deviation of vegetation period until flowering (days)		18.46			

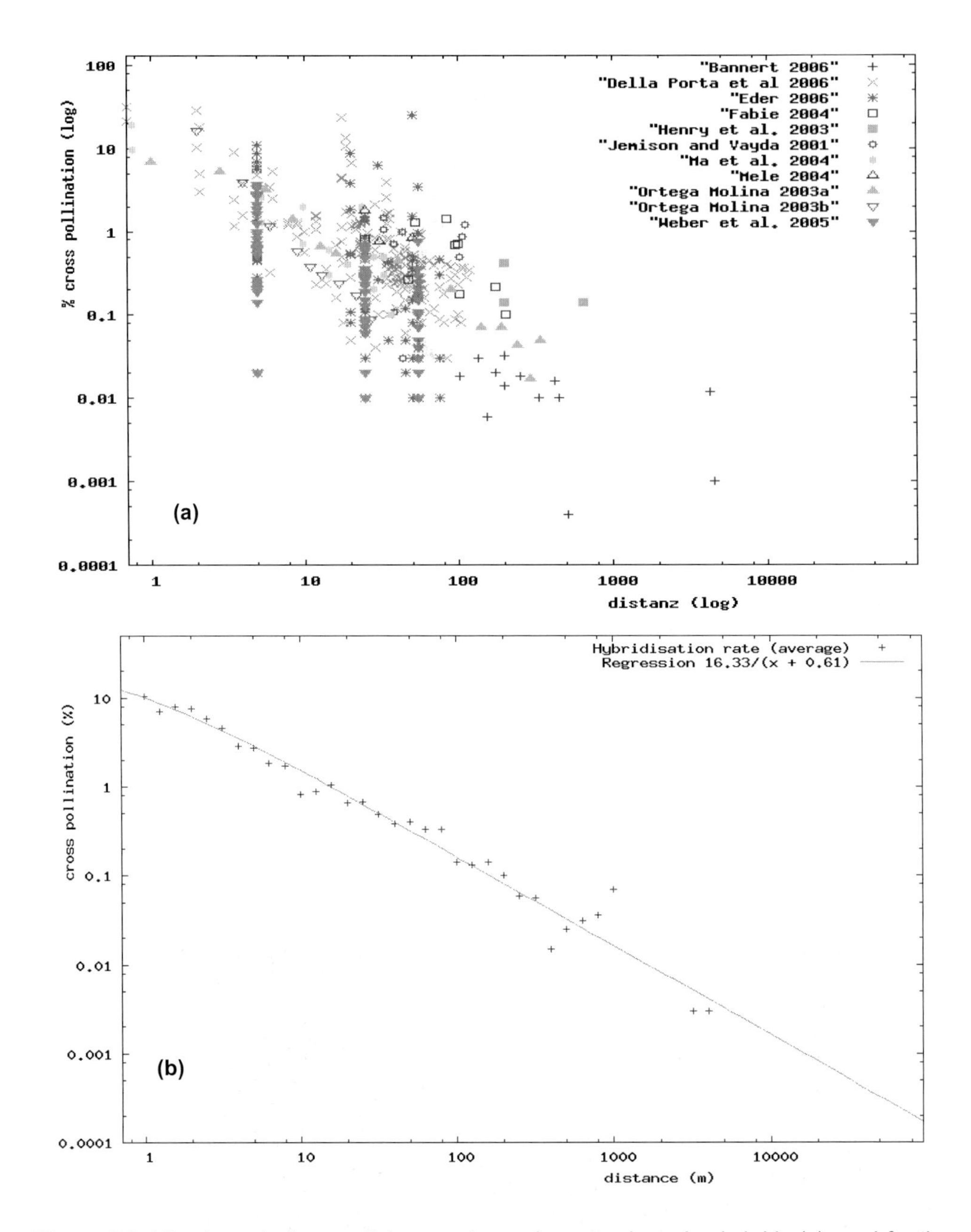

Figure 5.9: Literature studies on distance out-crossing rates in maize hybrids (a) used for the development of the dispersal kernel (b), projected on a double logarithmic scale. The data shows a regression function with a gradual reduction in hybridization rates from 0.14 % at 100 m, 0.1 % at 200 m, and 0.06 % at 250 m and about 0.01% at over 1000 m. The displayed references in (a) are quoted in the reference list. The data of Ortega Molina (2003a) and (2003b) were republished in 2004. Detailed description is provided by Reuter et al. (2008).

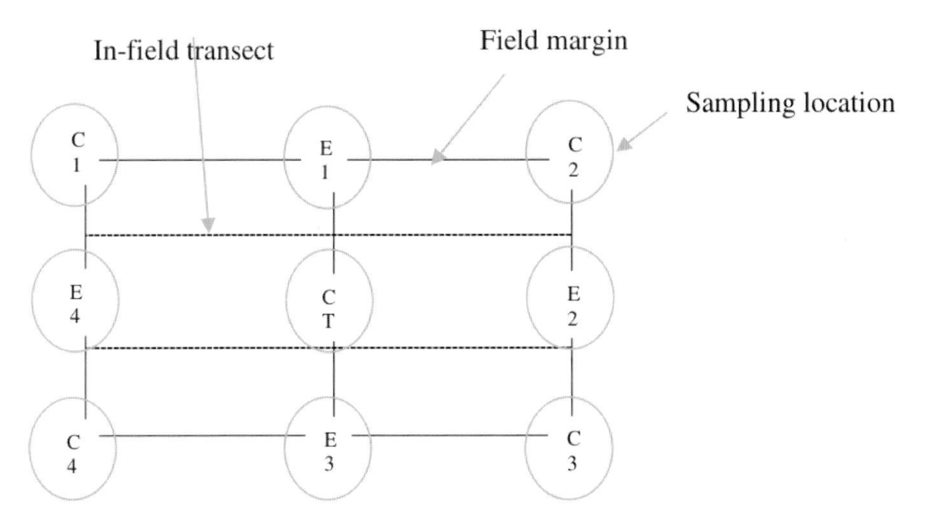

Figure 5.10: Diagram of sampling scheme used to estimate temporal heterogeneity of maize flowering and pollen exposure. Eighteen (18) farms were randomly selected for this study in order to determine the average time span during which pollen is available. Nine sampling locations were investigated per field, namely field corners C1, C2, C3, C4; field edges E1, E2, E3, E4; and center CT marked along transect gradients. Two plants were tagged per location and date recorded at 'time of first pollen availability' as well as date 'pollen shedding completed' (see Fig. 5.11 below). These were achieved through daily observations of all tagged plants. The duration of pollen availability was based on average value from all points.

Figure 5.11: Estimating the duration of pollen exposure through tagging procedure at the center of a field.

The data on the vegetative phase until flowering was experimentally determined with the help of sowing data obtained from farmers and a record of onset of flowering. Pollen shed data was estimated through tagging experiments recording the 'time of pollen availability' as well as the noting of the date 'pollen shedding' completed (see Fig. 5.11). Maximum distance for dispersal of pollen was based on potential distance by virtue of insect activities. For instance, bees could transport pollen up to a distance of about 5,000 m.

5.4.3 Map specification and investigated scenarios

Four (4) model scenarios were assessed, covering a range of different maize seed acquisition modes by farmers as obtained from the socioeconomic data. These data were subsequently transformed into an initial model map that provided a peculiar spatial pattern stipulating the exact locations where the farmers grow their seeds. These may be summarized as:

- Maize field locations (Gauss-Krüger coordinates of field locations);
- Maize area (in square meters and in hectares) for each location;
- Maize traits (a scenario specification whether a field was assumed to be grown with conventional crops or their genetically modified counterparts); and
- Field identification number.

For the scenario, the field location data and responses of the 201 farmers were used who participated in the socio-economic survey. This implies, for example, that a field assumed to be sown from food grains was actually managed that way. For the rest of the fields where no seed source data were actually available, it was always assumed in the scenarios that they were cultivated conventionally. The model was run 10 times per scenario and average calculations written to an output file. The repeated runs distribute pollen between all fields according to the variability range to derive quantitative estimations of pollination rates, taking into account the synchrony effects of the flowering phase. The individual scenarios analyzed include:

(a) Scenario 1: This assumed that GM seeds planted were obtained under farmer exchange conditions. This implied that the seeds were obtained from other farmers as gifts or simply exchanged. Here, an analysis involving 42 GM fields among 1,348 conventional fields is derived.

(b) Scenario 2: This assumed that GM seeds planted were obtained from the food market. This points to the fact that some farmers plant seeds from maize lots originally bought for food. Here, total number of GM fields planted is 35 occurring among 1,355 conventional fields is analysed.

(c) Scenario 3: An assumption that GM seeds planted obtained from the seed market and extension services implying the use of commercial varieties. In this simulation a total of 117 GM fields occurring among 1273 conventional fields are analysed.

(d) Scenario 4: Finally, we assumed the presence of a single GM field in the center of the study area. This suggests the scenario of a single GM field among 1,389 conventional fields.

5.4.4 Estimating duration of pollen availability

Already, the sampling approach for the estimation of the temporal heterogeneity of the flowering phase of maize populations was implemented in the field (Figure 5.11). Eighteen (18) seemingly rectangular fields occurring within a 1 km^2 area were selected for this study (Fig. 5.12). The exercise sought approval of respective farmers involved, which also allowed obtaining information of particular cultivars grown and their sources.

5.5 Socio-economic survey with smallholder farmers

Data was obtained from 201 farmers, involving the administration of formal questionnaires from May - November 2006 within the peri-urban location of 25 km^2 of Accra. Instructional sessions were organized for three selected enumerators prior to the interviewing exercises. All three enumerators had agricultural science backgrounds, thus had familiarity with the issues of interest. In many cases, interviews were conducted jointly by researcher and one of the enumerators; with one interviewing and the other recording emerging responses. On the average, an interview lasted between 40 minutes to about 1.30 hours, depending on farmer's cooperation and extent of farm activities being carried out. It was hardly possible to locate a respondent on the first visit. Hence enumerators spent about 2-3 days on average at certain sites to complete questionnaires. In situations where farmers could not be located after repeated visits, other farmers were chosen at random. Interviews followed a simple random sampling approach[2]. Following a day's interview, questionnaires (Appendix 2.2) were reviewed for completeness and assigned a number code. The questionnaire had the following details:

(a) Introductory part: Here, the context of the study, the objectives and the general criteria were explained. Relevance of aspects of their cultivation practices, crop production, including crop history, rotation measures, seed origins, weed and pest management, and economy factors were explained to the farmers.

(b) General information: This section documented the initial data:

- Name of respondent (Farmer)
- Name of investigator
- Date of farmer interview
- Farmland location on map (GPS values)
- Date farmland was mapped

---

2 In a simple random sampling procedure, each member of a population has an equal chance of being selected (sampled) and each combination of members of the population has an equal chance of making up the sample (Lohr 1999).

- Name of Community (Urban suburb)
- Questionnaire code (Questionnaire identification number for the respective day, corresponding with the name of person who conducted the interview).

(c) Respondent's profile: This section documented the following data:
- Age and sex of respondent
- Marital status
- Place of origin
- Nr. of children
- Nr. of persons in household
- Educational status
- Main occupation
- Per capita monthly income (estimate in US$)
- Per monthly household income (estimate in US$)
- Period of stay in the community (years):

(d) Agronomic practices and farm economy: here, farm activity profile, land use history, labour requirements, and land tenure types were documented (Table 5.10). In addition, the section was used to document components and costs of farm investments and expenditure among others.

(e) Crop management issues: An overview of this section is shown in Table 5.11.

Table 5.10: Agricultural practices and smallholder economy.

Variable	Codes/ Descriptors
Maize cropping intensity:	[1] 2x/year maize cropped; [2] 2x/year maize intercropped; [3] 1x/year maize cropped fallow; [4] 1x/year intercropped e.g. tuber crops; [5] 1x/year intercropped e.g. non-tuber crops; [6] No response
Land tenure:	Land use regime: [1] Private land; [2] Leasehold; [3] Public/Usufruct
Land rights (Specific rights to cultivation):	[1] Owner; [2] Care taker; [0] No rights
If leasehold, how many years?	Number years land is leased.
Number of fields (or plots) owned:	Number of fields owned by respondent.
Farm type:	[1] Subsistence/ for family supply e.g. back- or front-yard gardens, public areas, uncompleted buildings, within compounds, etc.; [2] Commercial/ for market supply e.g. These are found in open areas, also public places, along highways etc.
Labour requirements:	[1] Family; [2] Hired; [3] Both.
How many persons engaged on-farm:	Number of farm hands, if any

Table 5.11: Crop management conditions (farm site-specific).

Variable	Code/ descriptors
Time of planting and harvest:	Referring to the month
Duration of cultivation:	Estimated in weeks
Sources of seeds:	Seed acquisition from: [1] Previous harvest; [2] Formal seed system/shops; [3] Gifts e.g. from other farmers; [4] Seeds from extension agents.
Criteria for seed selection:	[1] High yielding; [2] Stable varieties; [3] Insect resistant; [4] Low input costs; [5] Early maturing; [6] Gift; [7] Other
Name of varieties planted:	Specific name of maize varieties sowed
If seeds were bought, are hybrids re-planted?:	[1] Yes; [2] No
Reasons for (-or) not re-planting hybrids	
Weed control measure:	[1] Use of cutlass; [2] Hoe; [3] Tractor; [4] Other
If non-tillage, specify herbicide brand applied:	Name of herbicide used.
Soil fertility management:	[1] Green manure; [2] Mulch; [3] Compost; [4] Inorganic fertilizers; [5] Nothing.
Limiting factors to productivity:	[1] Rainfall variability; [2] Insect pests [3] Wild animals; [3] Domestic animals; [5] Parasitic weed [6] Nutrient deficiency; [7] Diseases/ fungal pathogens, streak viruses etc; [8] Erosion; [9] Other
Agricultural information acquisition:	[1] Farmer-Farmer; [2] Extension Services; [3] Media; [4] Research Project; [5] Others
Evaluation of potential harm of a potential cultivation of GM crops	[1] Stable, lower-yielding variety from local inexpensive source; [2] Initially high yielding from the formal seed system, requiring higher inputs, more expensive with declining yields (as the assumption with Bt crops with which pests develop resistance over time.
Existing forms of informal or volunteer cooperation:	[1] No; [2] Yes

5.6 Molecular genetic methods

5.6.1 Plant materials

Sampling of maize seeds (*Zea mays*) involved the collection of locally grown seeds from 60 farmers within the study area in 2006. Protocol to document known commercial varieties as developed by the GGDP is listed in Appendix 1.2. Available commercial varieties were purchased from the formal seed system, mainly shops within the localities shown in Table 5.12. The shops in the localities of Amasaman and Nsawam are commercial enterprises largely patronised by local people, while the location in Accra was chosen at random. Two additional varieties from Seville (Spain) and an exotic variety bought in Bremen (Germany) were assessed synchronously with the local samples for analysis of species pedigree. Details of the two commercial cultivars analyzed are presented in Table 5.13.

5.6.2 Preservation and storage

A collection of 20-30 individual maize seed samples per population were treated with Actellic dust (composition: Pirimyphos-methyl and Permethrine), stored in small plastic bags and labeled (Fig. 5.15). The seeds were subsequently transported in January, 2007 under an import permission issued in Germany - Reference No. 53-18, Der Senator fuer Arbeit, Frauen, Gesundheit, Jugend und Soziales, Freie Hansestadt Bremen, 11.01.2007 - for further storage in the laboratories of the UFT, University of Bremen (Storage conditions: 4°C in cold room; 20. January– 15 December, 2007).

5.6.3 Experimental steps

(a) DNA extraction: The extraction of total genomic DNA was conducted in the laboratories of the UFT, University of Bremen in January 2008, following the implementation of test steps to assess the feasibility of extraction from fresh leaves and seeds between November and December 2007. For the main experiments, DNA was extracted directly from seeds owing to very low recorded germination rates obtained for all the planted samples. Two extraction methods namely DNeasy® and the Chelex® methods were directly applied in relation to the seed samples, to be able to select a method that provides the most reliable and optimum yield of DNA.

Table 5.12: Procurement locations of commercial varieties.

Name of cultivar (Species: *Zea mays*)	Location of shop
Obaatanpa	Amasaman
Obaatanpa	Accra (main township)
Obaatanpa	Nsawam
Mamaba	Accra (main township)

Table 5.13: Commercial maize hybrids analysed.

Name (Year of release)	Grain colour	Grain texture	Maturity- days to flowering)	Yield (t/ha)	Streak resistant?	Nutritionally enhanced?	CIMMYT Germplasm
Obaatanpa (1992)	White	Dent	105	4.6	Yes	Yes	Pop 63-SR
Mamaba (1996)	White	Flint	110	6.0	Yes	Yes	Pop.62, Pop.63-SR

Source: Morris et al. (1999) (These are varieties developed by the Ghana Grains Development Project (GGDP) in collaboration with CIMMYT. The list of maize and hybrid varieties developed under the GDDP is documented in Appendix 1.2.

The DNeasy® method: This was used following the manufacturer's instructions[3].

The Chelex® method: The Chelex resin is composed of styrene divinylbenzene co-polymers containing paired iminodiacetate ions, which act as chelating groups (Walsh et al. 1991). The Chelex® method was selected based on DNA yield data obtained from 23 DNA extracts from each method.Twelve individual seeds[4] were pooled from each population, washed clean from preservative, and placed in small plastic bags. The samples were then crushed with a hammer and ground into fine powder in a mortar with the help of liquid nitrogen. The following experimental steps were subsequently followed:

- 450mg powder of each sample was weighed out;
- Addition of 200µl of 5% Chelex® solution (presence of Chelex during boiling prevents degradation of DNA by chelating metal ions that may act as catalysts in the breakdown of DNA at high temperatures in low ionic strength solutions (Walsh, 1991)
- Addition of 10µl 100mM DTT (to destroy the secondary structures of the DNA);
- Addition of 8.0µl of 20mg/ml Proteinase K (Denaturing the proteins);
- Incubation in a thermoshaker at 54°C for 3.5 hours and1000 rpm;
- Heating for 95°C for 5 mins (Denaturation step of Proteinase K);
- Centrifugation at 13,400 rpm for 5 mins.
- Pipetting of resulting supernatant (i.e. extract of total genomic DNA).

(b) DNA yield estimation: DNA yields and purity estimations were obtained based on absorbance measurements with a spectrophotometer (NanoDrop, Version 3.2.1, 1000; Peqlab Biotechnology GmbH, Erlangen).

(c) PCR amplification and SSRs: Two fingerprinting techniques - RAPD (Random Amplified Polymorphic DNA) and SSR (Simple Sequence Repeats) were tested for their suitability in distinguishing small genetic differences between amplified fragments, and to confirm possible marker sets selected. Subsequently, SSR markers were chosen following higher polymorphism shown in preliminary experiments, and the feasibility to distinguish between fragments. A set of 4 primer pairs utilised are tabulated in 5.14. Primer selection (i.e. phi108411, phi331888 and phi96100) were based on recommendations of Dubreuil et al. (2006) and Warburton et al. (2002); and phi031 based on Dubreuil et al. (2006).

In all, 50µl PCR reaction mixture was prepared into small Eppendorf cups containing 46µl of master solution (see Appendix 4.5.2), and 4µl of genomic DNA template. Thermocycling (iCycler, BioRad, Hercules, USA) programme involved the following steps: 1) One cycle (initial denaturing step) at 94°C for 10 min; 2) 30 cycles at: 94°C for 30 seconds (denaturing); 40°C for 1 min (annealing); and 72°C for 1 min (extension); 3) One cycle at 72°C for 5 min (final extension); and 4) One cycle at 4°C (storage condi-

---

3 DNeasy® Plant Mini Kit- For miniprep purification of total cellular DNA from plant cells and tissues, or fungi, DNeasy® Plant Handbook. July, 2006. www.qiagen.com

4 Followed recommendations of Warburton, M. (CIMMYT) in personal communication in January, 2008, as being an adequate representation of single populations.

tions). The documentation of reagents, materials and instruments used are presented in Appendix 3.

(d) Electrophoresis: A 2% agarose gel was prepared for DNA gel electrophoresis. 5 µl of PCR amplification products diluted by mixing with 5 µl loading dye on parafilm (obtained from Fermentas, St. Leon Rot). Alongside 10µl of ladder (Mass Ruler™ DNA ladder of 10,000-80 bp, St. Leon Rot), 10µl each of all PCR products were pipetted into 28 comb wells created within self-cast gels prepared in 27cm long electrophoretic chamber. Electrophoresis runs were done at 90V constant power for 4 hours. Gels (containing DNA band fragments) were placed in prepared solution of ethidium bromide for staining (5x10-5%), and subsequently placed on a shaker (TH15, Edmund Bühler, Tübingen) for 1 hour. The gels were illuminated and the resulting bands photographed with a UV transilluminator (Vilbert Lourmad, Mame-la-Vallee, France).

Table 5.14: SSR Primer sets used. Primers were all obtained from MWG Biotech Company.

SSR Primer name	SSR Primer sequence
phi108411F	CgTCCCTTggATTTCgAC
phi108411R	CgTACgggACCTgTCAACAA
phi331888F	TTgCgCAAgTTTgTAgCTg
phi331888R	ACTgAACCgCATgCCAAC
phi96100F	AggAggACCCCAACTCCTg
phi96100R	TTgCACgAgCCATCgTAT
phi031F	gCAACAggTTACATgAgCTgACgA
phi031R	CCAgCgTgCTgTTCCAgTAgTT

5.7 Statistical evaluation and softwares used

Using the programming language SIMULA, specific procedures were written to find out the nearest neighbour relations, and the number of field neighbours within a certain distance interval. These were the Nearest Neighbour Analyses (NENA), a program that calculates distances to nearest neighbour field by comparing 2 data sets on field locations, one for origin, one for neighbour; and Number of Neighbour Analyses (NUNA) which counts the number of field neighbours within certain distance ranges. Spatial analysis and the calculation of farm acreages and overlay of ground GPS data on Google satellite imagery was achieved with the help of ArC GIS application.

The Maize Model (MaMo) was used to simulate regional cross-pollination between GM and conventional fields through a dispersal kernel (function). It helped to calculate pollen hybridization rates (%) in recipient fields and provided estimates on average presence of GM in conventional harvests, and the average conventional presence in GM fields. STATISTICA Version 6 (Statsoft Inc. 2001) and MS EXCEL (2003) supported the evaluation of the socioeconomic interview data. For the genomic analysis, DNA yield and purity estimations were done with the help of the NanoDrop (Version 3.2.1,

1000; Peqlab Biotechnology GmbH, Erlangen). For checking reproducibility of bands, TotalLab software (Version 2.01, Newcastle, UK) was used. Comparison of band lengths was possible with the use of DNA ladder system O'RangeRulerTM of 300-20 bp and Mass Ruler 10,000- 80 bp from Fermentas.

Chapter 6: Results

This section presents the findings of the study. We focus initially on the geographic results and spatial evaluation followed by the cropping demographic aspects. The model results are then presented and subsequently followed by the socioeconomic aspects. The genomic findings are then finally presented.

6.1 Geographic Aspects

Spatial analysis of maize cultivation covered a coherent 25 km^2 area of Accra West, and identified a wide extent of household gardening and open space farming - both regarded in this context as farms/ fields - as well as feral stands found within the same period as specified in Table 6.1.

6.1.1 Nearest neighbour distance analyses

Analysis of nearest neighbour of farm locations reveal the following (Fig. 6.1.1). About 98.8% of all farms lie within nearest distance range of 5-150 m. Shortest nearest distance between farm plots is 5m with a frequency of about 60 farm plots. Largest nearest distance between farm plots within the area of 25 km^2 is 459 m. Two modal peaks are observed with a frequency occurrence of 130 farms with nearest distances of 30m and 45m respectively.

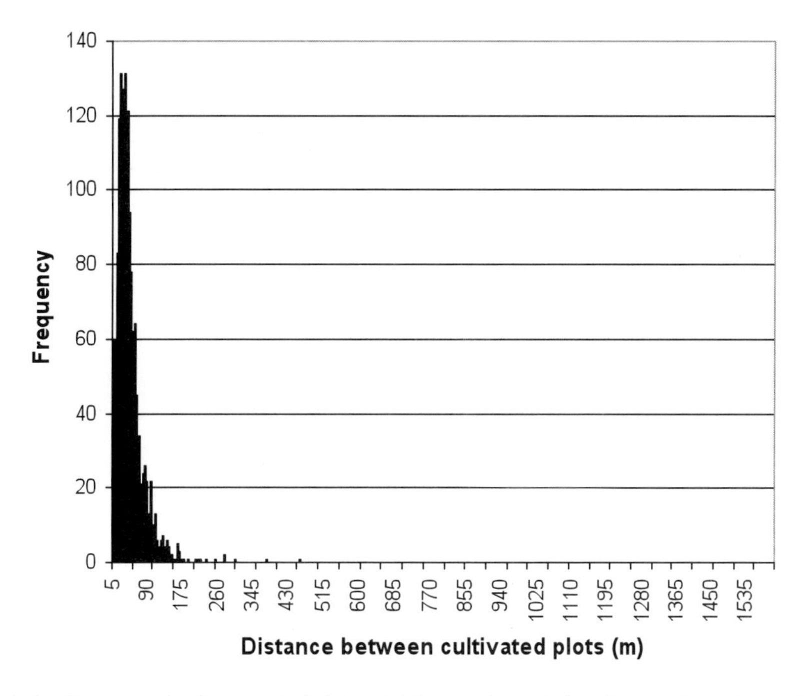

Figure 6.1.1: Distance to the next field neighbours for all locations of maize cultivation (n=1,390) on a scale of 5 m distances.

Table 6.1: Characteristics of the study area.

Item No.	Item	Description
1	Study area	Ga West District of Accra. Ghana
2	Latitude, Longitude	5° 42' 0N, 0° 18' 0W
3	Settlement description	Urban periphery
4	Population size	348,926 with female population representing 49.9% of the total population; males make up the other 50.1% (National Population and Housing Census, 2000) with intercesal growth rate of 3.4%. The projected population for the year 2006 is 426,439 (Ghanadistricts, 2007)
5	Year of measurement	2006 (major rainy season from May - September)
6	Total area covered by investigation (km^2)	25.0
7	Number of sub-communities covered	13.0
8	Major arable crops of the region	Maize, cassava, beans, groundnuts and vegetables
9	Main types of arable cropping	Household gardening and open-space farming
10	Total number of maize stands (farms and feral locations) found	1,459.0
11	Total maize stand density (Km^{-2})	58.0
12	Total number of maize farms found	1,390.0
13	Cropping density (Km^{-2})	56.0
14	Number of feral locations found	69.0
15	Feral population density (Km^{-2})	3.0
16	Total maize area calculated from GIS records (Km^2)	1.1
17	Maize fractional area as a % of total study area	4.5%
18	Mean farm acreage m^2 (min-max) from GIS records	808.7 (1.0-35,132.0), with about 98% of all fields below 5000 m2 (0.5 ha)
19	Maize sites	Nine (9) farming site types identified are- Home compounds, traffic areas, construction sites, publicly accessible areas, privately-owned open spaces, small business areas; marginal sites, wetland areas and industrial places.
20	Nearest field distance (m)	98% of all field locations found are within nearest distance of 5-150 m.

In Table 6.2, statistics of farm cases within a distance range of 20 m up to a maximum of 200 m is presented.

Table 6.2: Number of farm neighbours within certain distance ranges (<200 m) for the area of 25 km^2 of Accra West. The minimum number of farm neighbours for all respective distance ranges is zero.

Distance ranges (m)	Mean number of farm neighbours	Maximum number of farm neighbours	Standard deviation	Standard error
0-20	0.2	3.0	0.5	0.01
20-40	0.6	4.0	0.8	0.02
40-60	0.8	5.0	1.0	0.03
60-80	1.0	7.0	1.2	0.03
80-100	1.1	8.0	1.2	0.03
100-120	1.4	11.0	1.4	0.04
120-140	1.4	10.0	1.5	0.04
140-160	1.7	12.0	1.7	0.05
160-180	1.7	13.0	1.7	0.04
180-200	1.9	12.0	1.8	0.05

In Fig. 6.1.2, it is observed that feral occurrences are more widespread, and in isolated numbers on the landscape when compared to farm data.

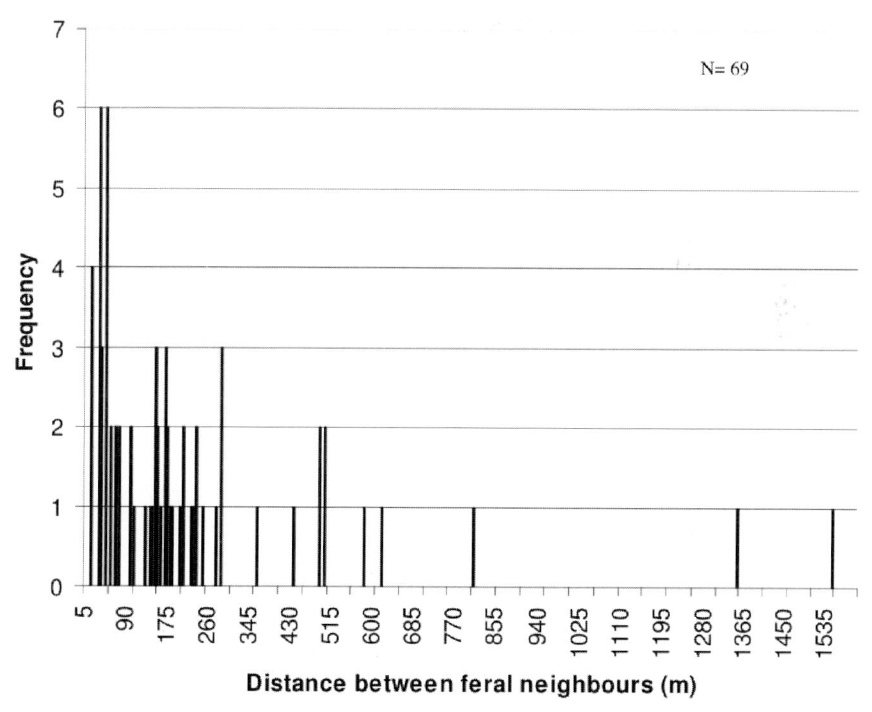

Figure 6.1.2: Distance to next feral neighbour for all locations of feral occurences on a scale of 5 m distances. Total number of feral stands found is 69 but more widespread across the landscape compared to farms with a feral stand density of 3 km-2. Shortest nearest distance recorded was 16m, with maize stand frequency of 4. Maximum nearest distance was 1,548 m with frequency ocurrence of 1 maize stand. Again, two modal peaks were observed with nearest distance of 36 and 49 m respectively.

The Table 6.3 presents the statistics of feral cases within a distance range of 20 m up to a maximum of 200 m.

Table 6.3: Number of feral neighbours within certain distance ranges (<200 m) for the area of 25km^2 of Accra West. The minimum number of feral neighbours for all cases was found to be zero.

Distance ranges (m)	Mean number of feral neighbours	Maximum number of feral neighbours	Standard deviation	Standard error
0-20	0.1	1.0	0.24	0.03
20-40	0.2	2.0	0.42	0.05
40-60	0.1	1.0	0.35	0.04
60-80	0.1	1.0	0.24	0.03
80-100	0.1	1.0	0.24	0.03
100-120	0.1	2.0	0.29	0.04
120-140	0.1	1.0	0.32	0.04
140-160	0.3	2.0	0.53	0.06
160-180	0.2	1.0	0.38	0.05
180-200	0.2	2.0	0.52	0.06

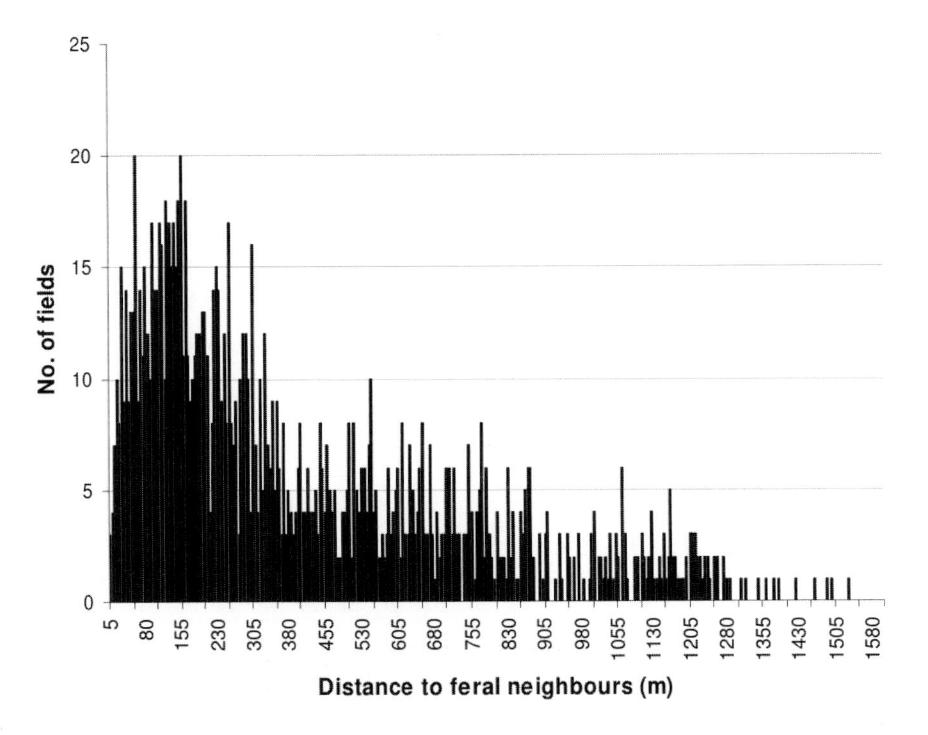

Figure 6.1.3: Nearest distance from farm to feral neighbours on a scale of 5 m distances. The Figure shows that nearest distance from farms to feral stands was 5 m. Maximum nearest distance recorded is 1,525 m. The figure reveals two modal peaks of 20 fields observed at distances of 35 and 160 m respectively.

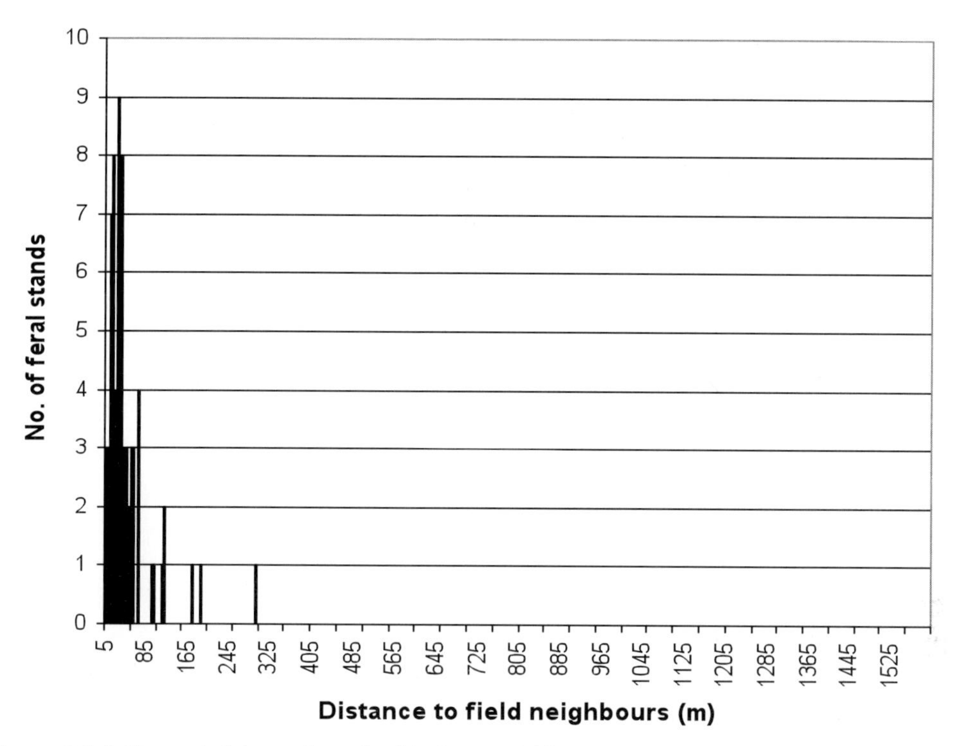

Figure 6.1.4: Nearest distance from feral to farm neighbours on a scale of 5 m distances. It depicts that the majority of feral stands (i.e. 89.9%) occur within 85 m from farms. Shortest nearest distance is 5 m; while the maximum nearest distance recorded is 297 m. The number of cases of feral stands was 69.

6.1.2 Spatial characterization of maize stands with the help of GIS

Fig. 6.1.5 shows the relative distribution of the maize crops on landscape. Potential habitation places are distinguishable. A correlation of these factors with a GIS operation is of special interest in this context since it provides information on precise locations where agricultural maize crops are planted and a reliable calculation of cultivation statistics e.g. mean density per km^2, feral demography and orientation of common farming areas. Surface types mainly roads, built public places, or main settlement areas mostly remain unchanged for longer time spans.

6.1.3 Spatial distribution of farms and ferals across larger distances

In order to derive estimations on the spatial variability of crop and feral occurrences across larger areas e.g. 144 km^2, an actual maize cultivation area (shown within the bold lined central block in Figs 6.1.6 a-c) with a size of 16 km^2 was replicated to cover 8 adjacent neighbouring blocks. The obtained pictures provide a basis for analysis basing on the assumption that the same pattern of cultivation distribution or feral occurences would be found also in the surrounding.

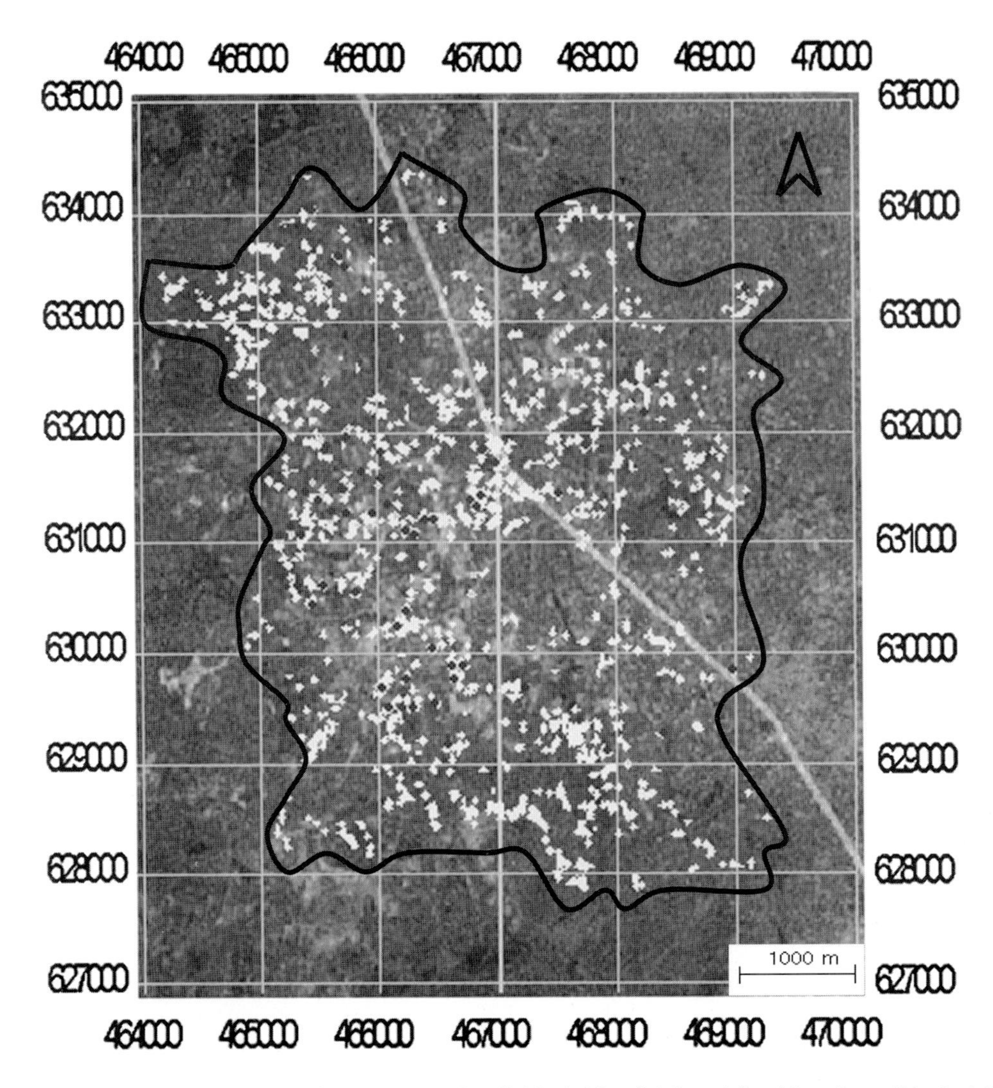

Figure 6.1.5: Distribution of maize cropping fields (white dots) and feral locations (black dots) on a 1 km2 grid in Accra West on Google image (http://maps.google.com) with a resolution of 1 x 1 km. The validation of ground cropping data with satellite data within the 25 km2 study area (marked blue) confirms a very dense cultivation practice across the landscape with almost any openly available space is utilised for cultivation.

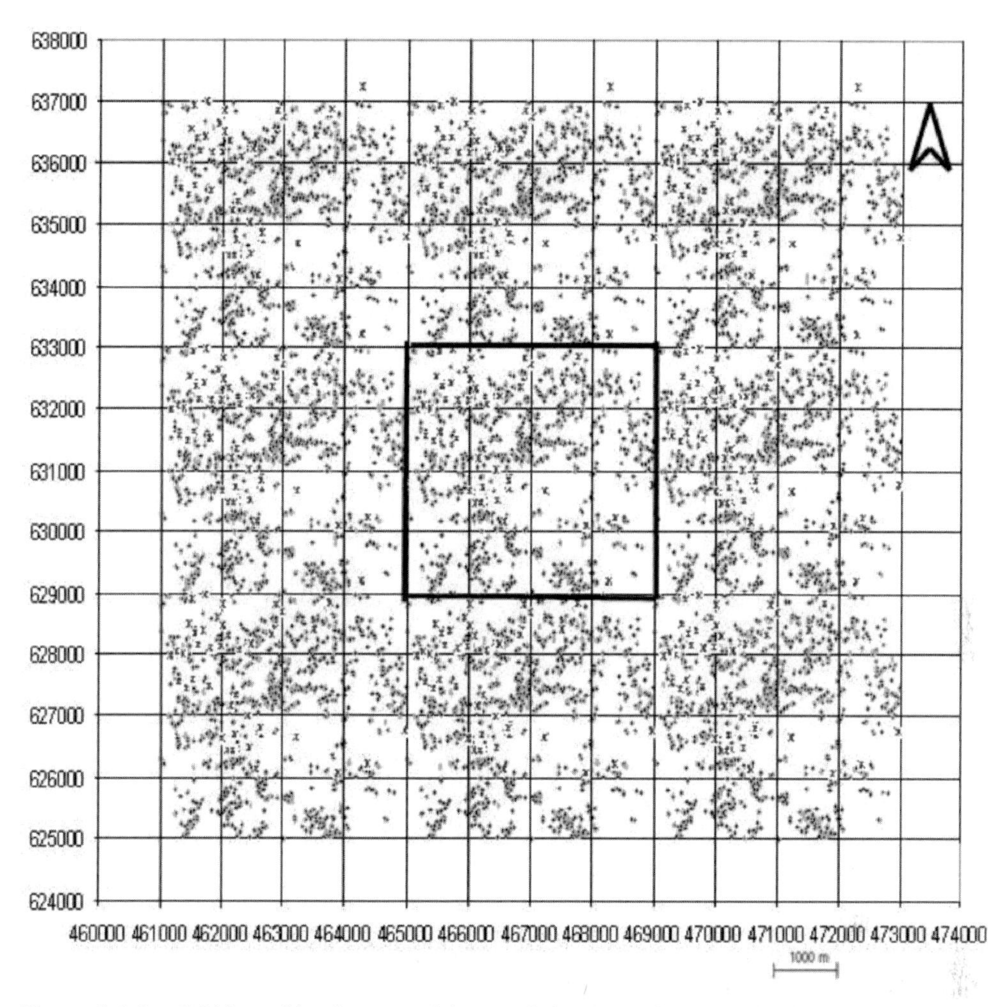

Figure 6.1.6a: Grid-based land use model to study land use due to maize field and feral stand occurrences within each 1 km^2 blocks for a 144 km^2 settlement region, obtained by replication of the central area.

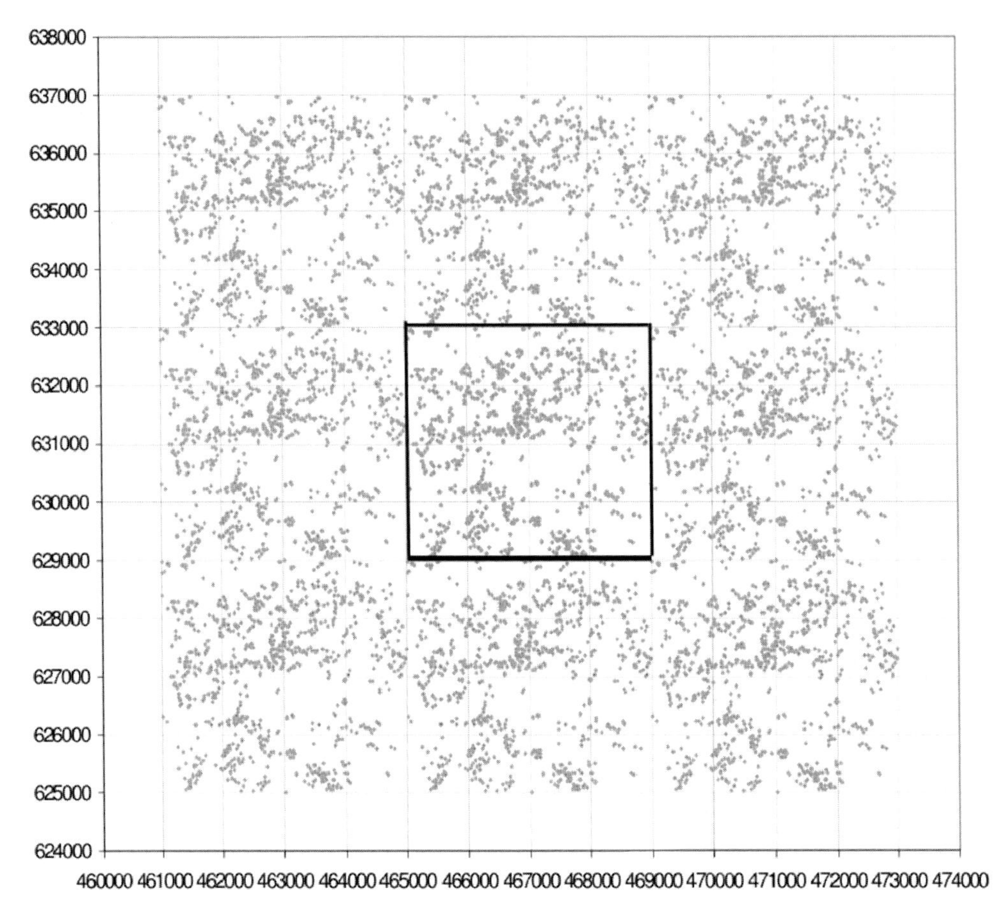

Figure 6.1.6b: Grid-based land use model to study land use due to maize field stand occurrences within each 1 km^2 blocks for a 144 km^2 settlement region, obtained by replication of the central area.

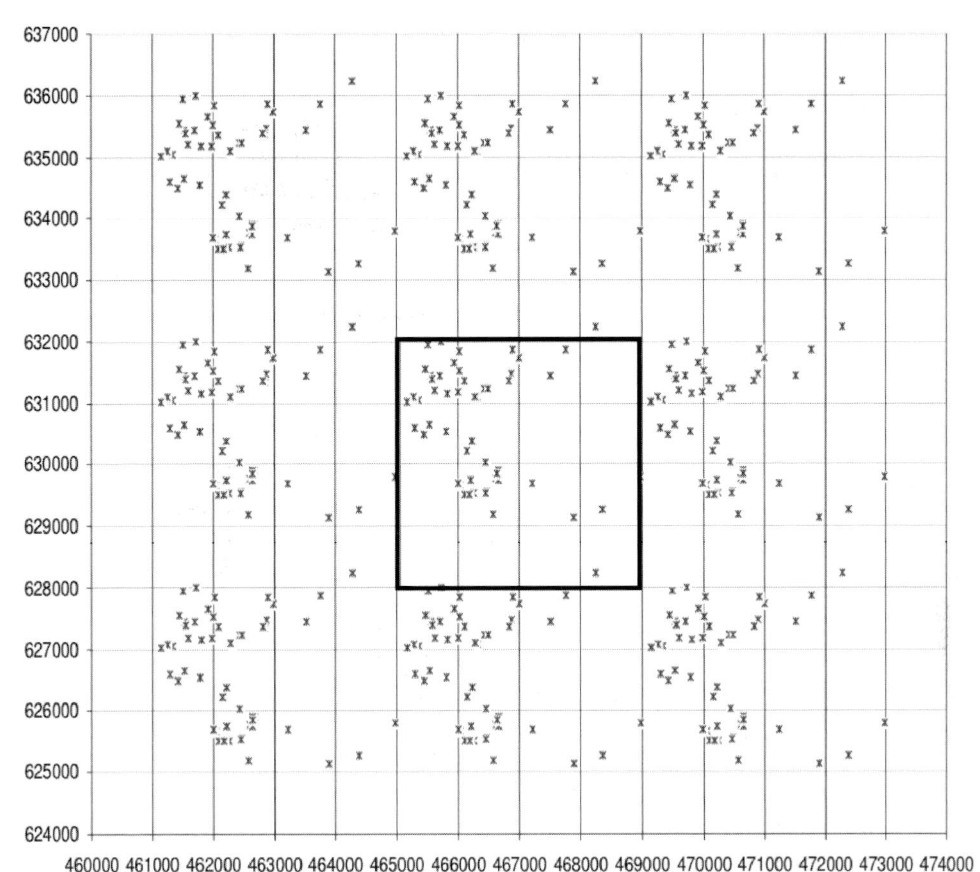

Figure 6.1.6c: Grid-based land use model to study land use due to maize feral stand occurrences within each 1 km^2 blocks for a 144 km^2 settlement region, obtained by replication of the central area.

6.1.4 Number of farm neighbours across larger distances

Basing on the distribution of farming locations as in Fig. 6.1.6 b, it was of interest to calculate the number of farm neighbours within the extended distances. Thus, in order of magnitude of 100 m distances, a classification of farm numbers from 0-100 m across to a distance of 3900-4000 m is presented in Fig. 6.1.7.

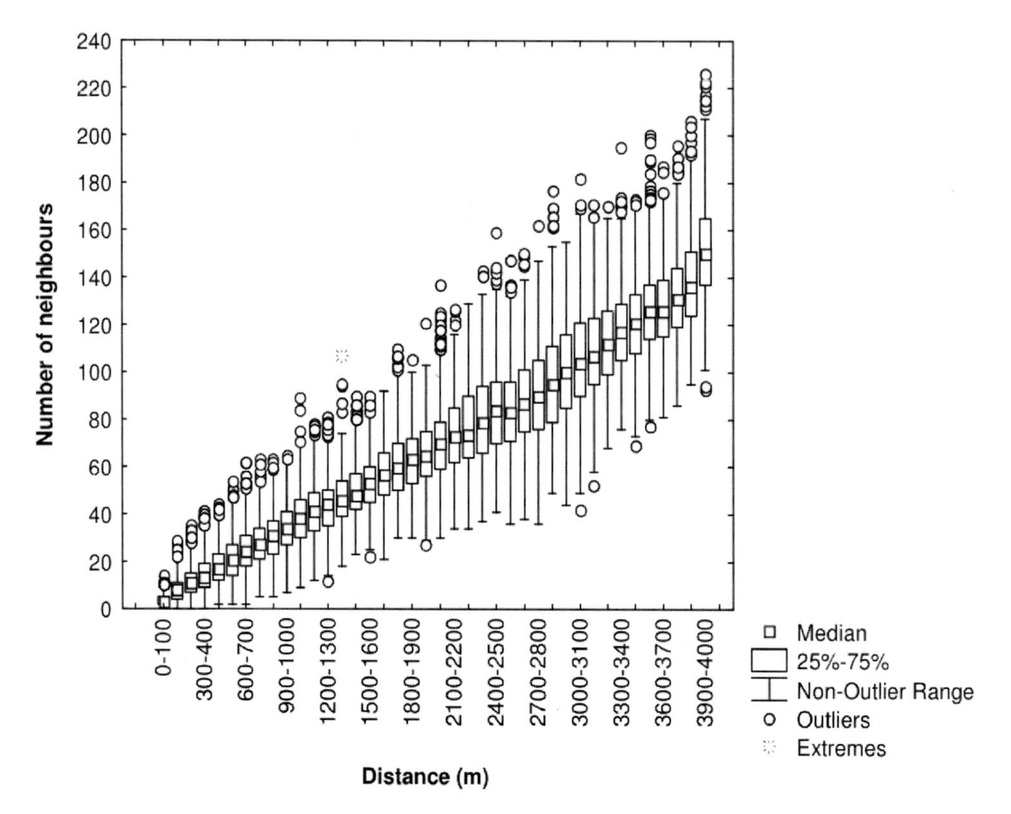

Figure 6.1.7: Number of farm neighbours within intermediate distance ranges in an area of 144 km2. Generally, a linear correlation is revealed between the numbers of cultivation locations with respective distances, allowing to calculate the cumulative number of neighbour cases in the region up to 4000 m. For example, there are a total of 4, 33, and 127 neighbours at 100 m, 500 m and 800 m distances respectively. For the intermediate distances there are about 4 farm neighbours on average within a distance of 0-100 m, increasing to a maximum value of 150 neighbours between distances of 3900-4000 m.

6.1.5 Number of feral neighbours across larger distances

In Fig. 6.1.8, a numerical classification of feral stands across larger distances is shown basing on data presented in Fig. 6.1.6.

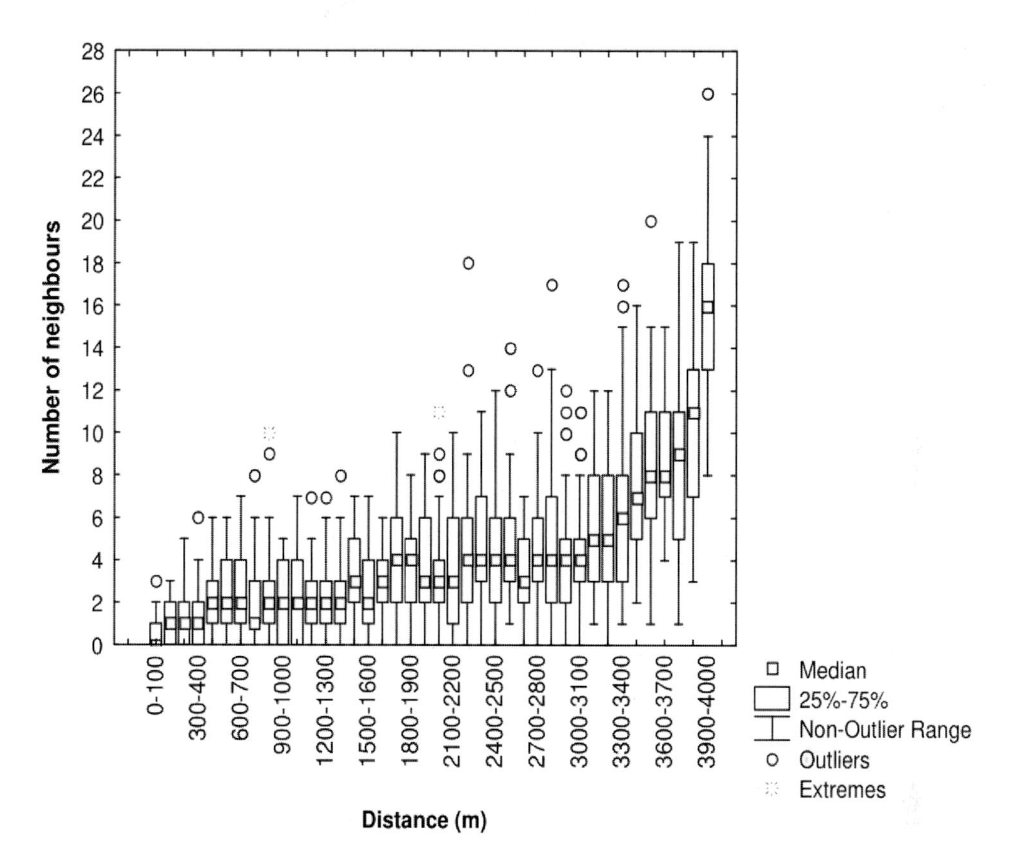

Figure 6.1.8: Number of feral neighbours within intermediate distance ranges in an area of 144 km^2. This also allows to calculate the cumulative number of neighbour cases in the region up to 4000 m. For example, there are a total of 4, 33, and 127 neighbours at 100 m, 500m and 800 m distances respectively. For the intermediate distances there are about 4 farm neighbours occur on average within a distance of 0-100 m, increasing to a maximum value of 150 neighbours between distances of 3900-4000 m. Since the feral occurrences are clumped (see Fig. 6.1.6 c), a non-linear relationship of number of neighbours and distance occurs.

6.1.6 GIS-based registration of maize field polygons

Size distribution is a crucial factor to assess the potential of gene transfer from fields to near natural habitats around the field. The interacting area involved depends not only on the distance proximity but also on the size and shape of fields within the neighbourhood. Figure 6.1.9 below presents the acreage coordinates by assigning GIS data to the corresponding regions (pixels) on the map of the study area. Further details are shown in Figs. 6.1.10 - 6.1.15.

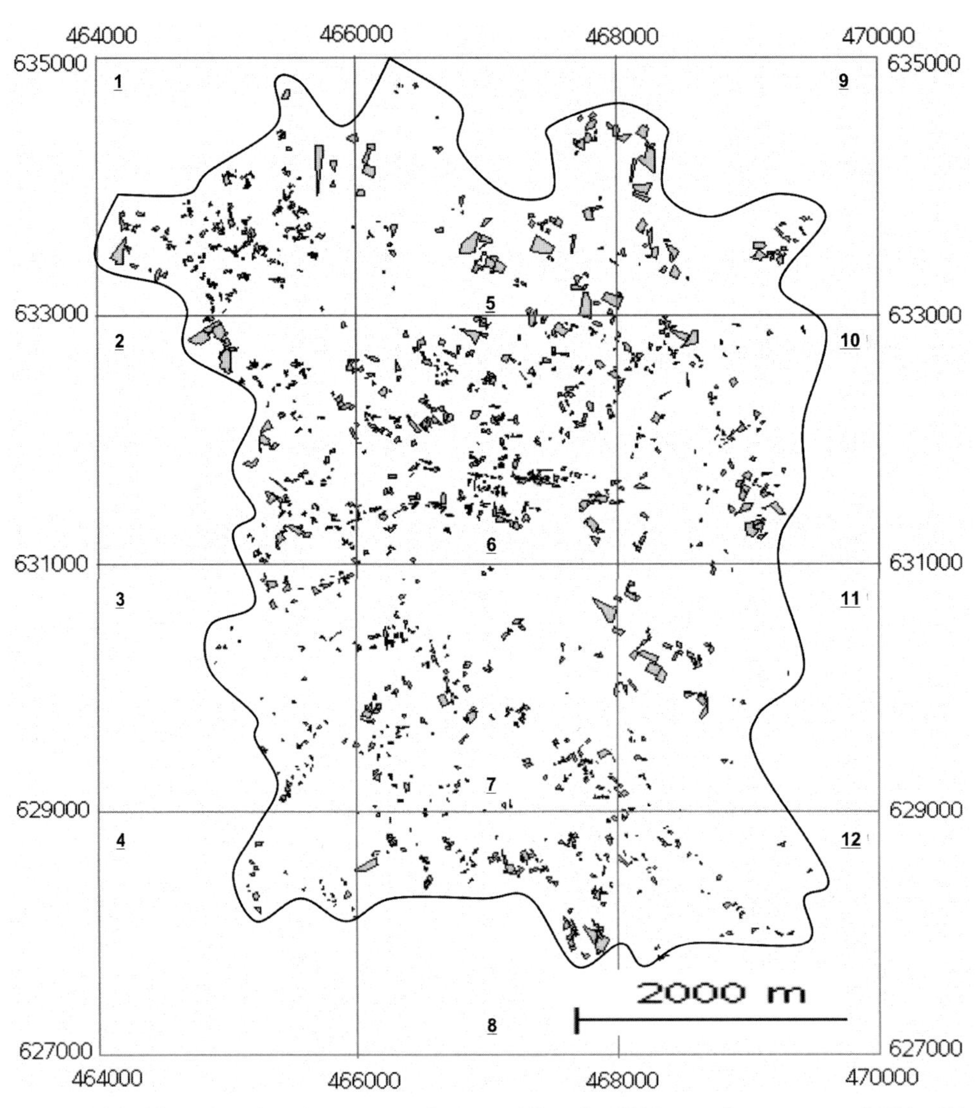

Figure 6.1.9: Map of total maize acreage in Accra West for 2006 covering an area of 25 km2 in twelve demarcated zones on a 2 x 2 km2 grid (WGS84, UTM Zone 30). The irregular dark line represents the boundary of the studied area. The border values are measured in meters.

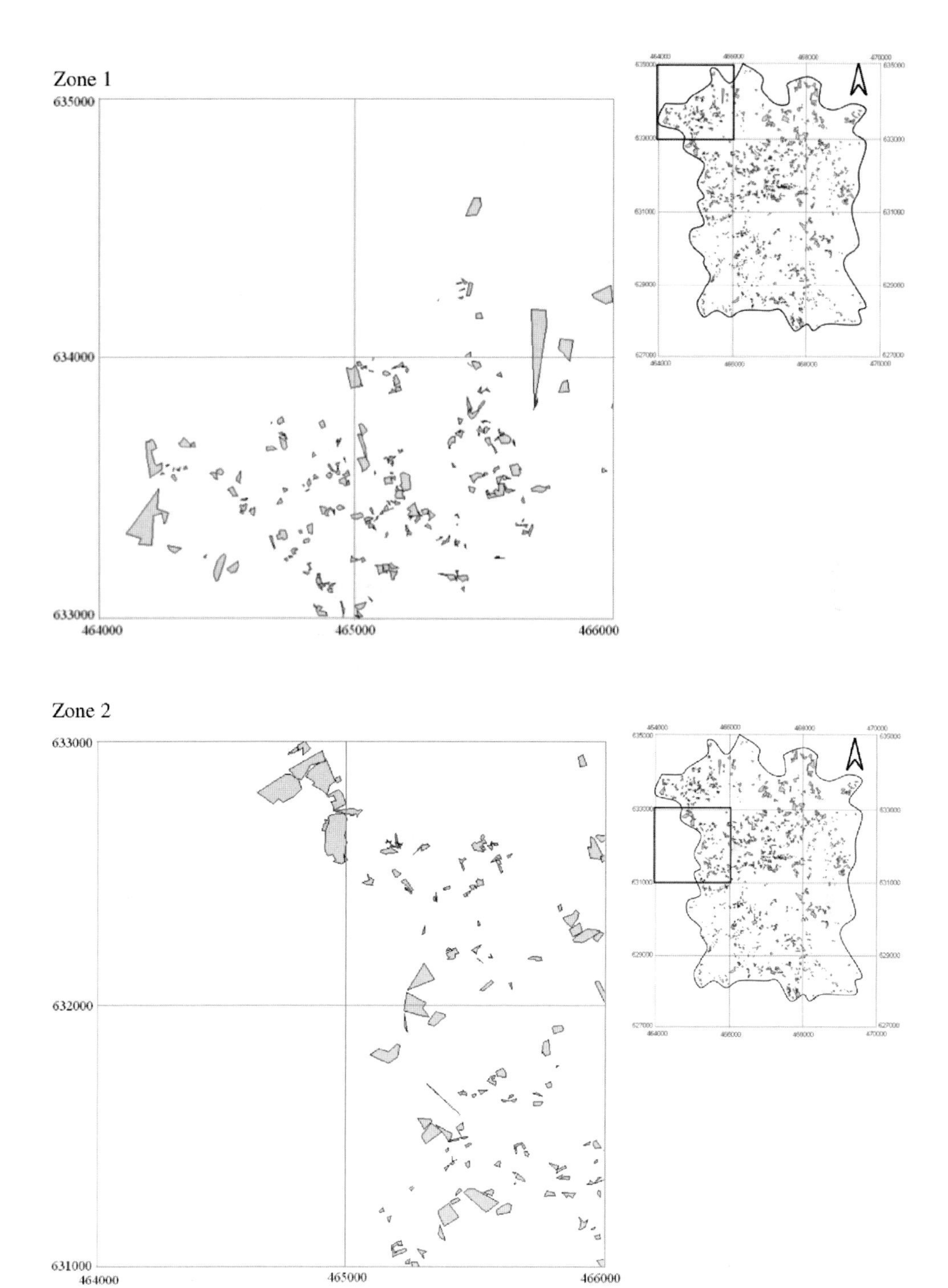

Figure 6.1.10: Spatial structure of maize fields in Zones 1 and 2. The relative positions of the frames are shown as inserts in the right top corner maps.

Zone 3

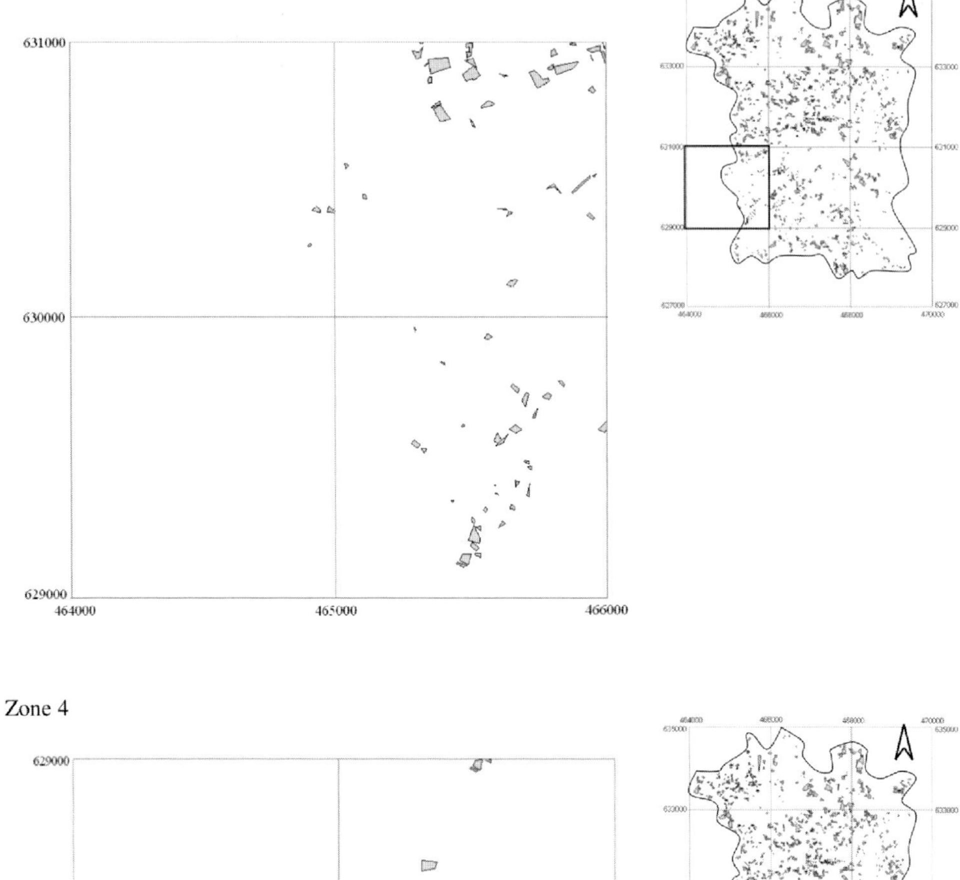

Zone 4

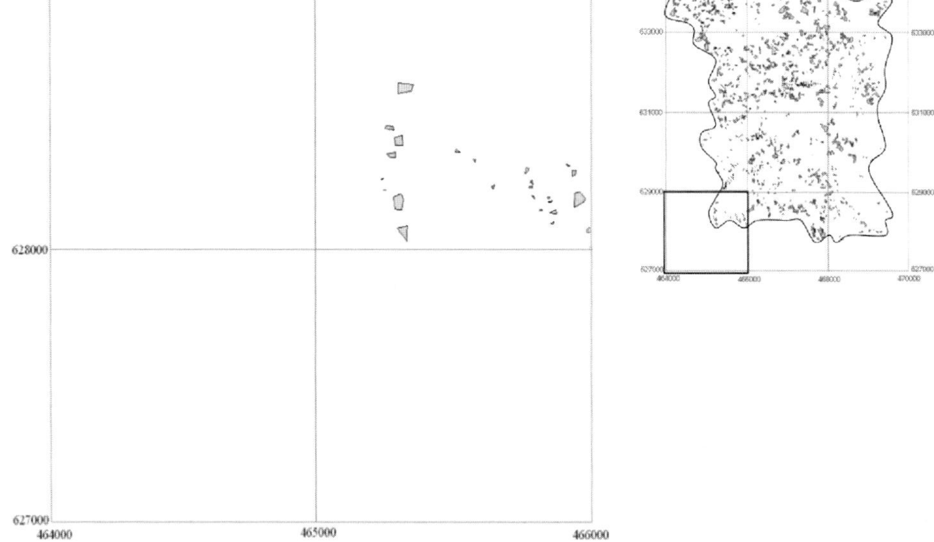

Figure 6.1.11: Spatial structure of maize fields in Zones 3 and 4. The relative positions of the frames are shown as inserts in the right top corner maps.

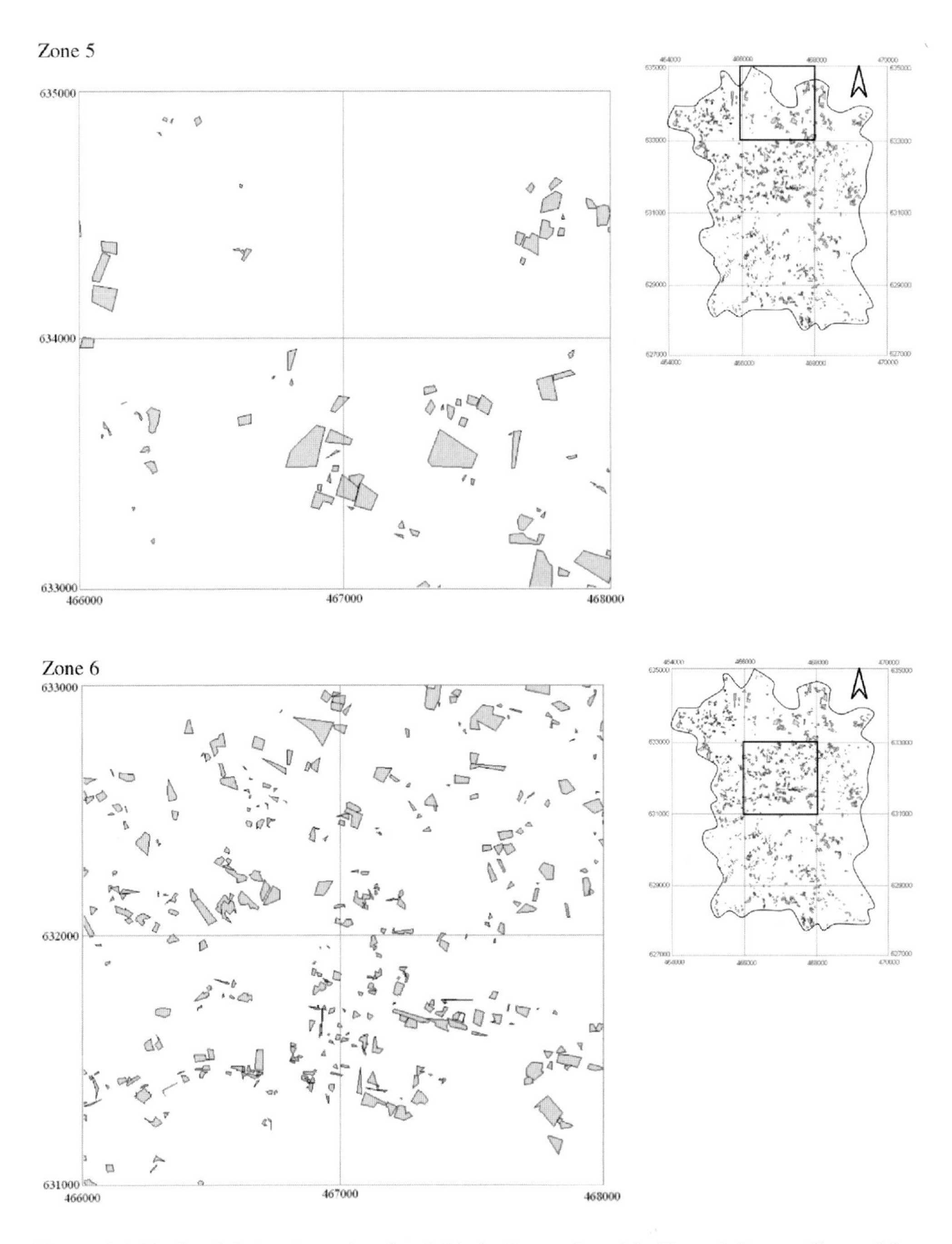

Figure 6.1.12: Spatial structure of maize fields in Zones 5 and 6. The relative positions of the frames are shown as inserts in the right top corner maps.

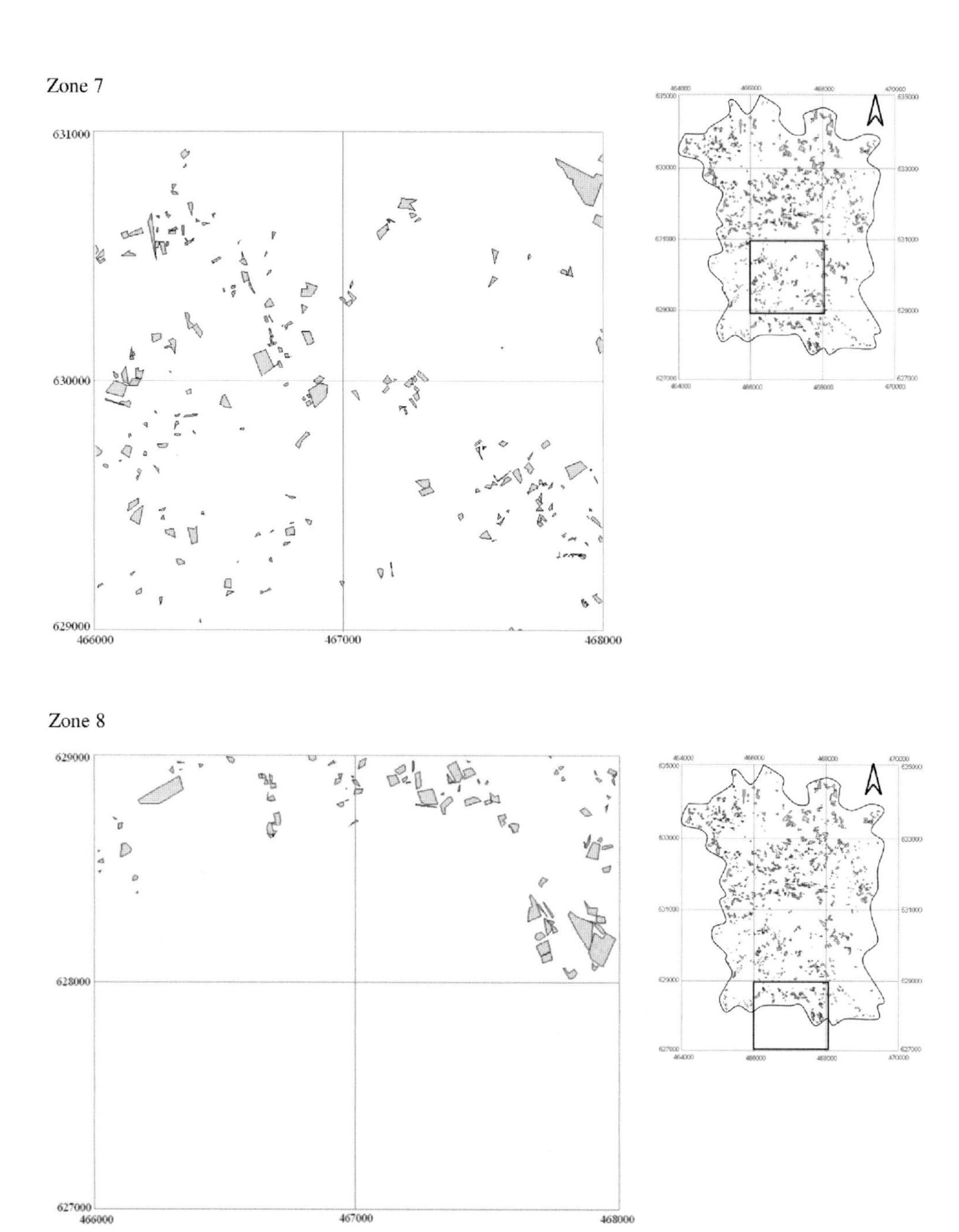

Figure 6.1.13: Spatial structure of maize fields in Zones 7 and 8. The relative positions of the frames are shown as inserts in the right top corner maps.

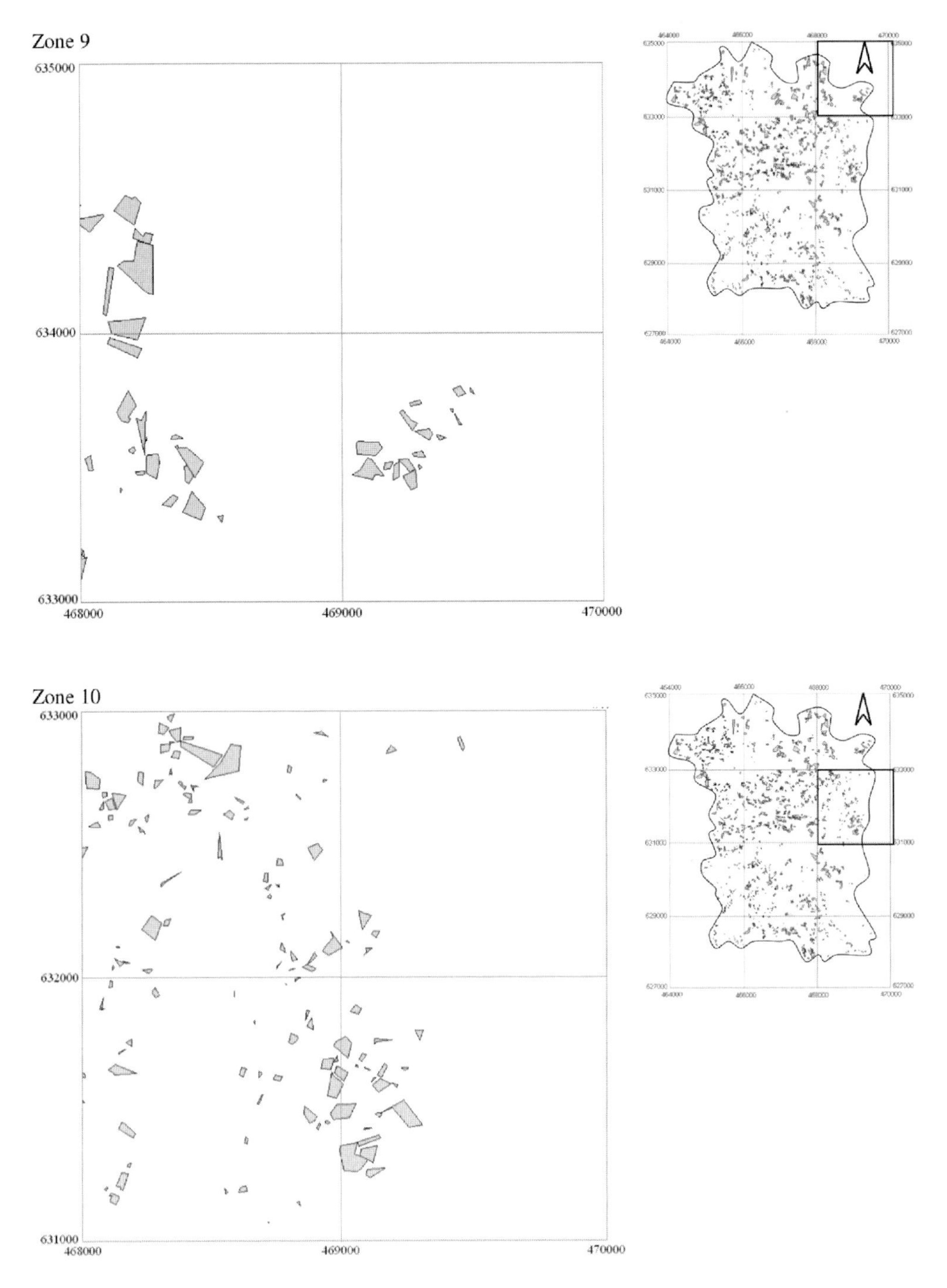

Figure 6.1.14: Spatial structure of maize fields in Zones 9 and 10. The relative positions of the frames are shown as inserts in the right top corner maps.

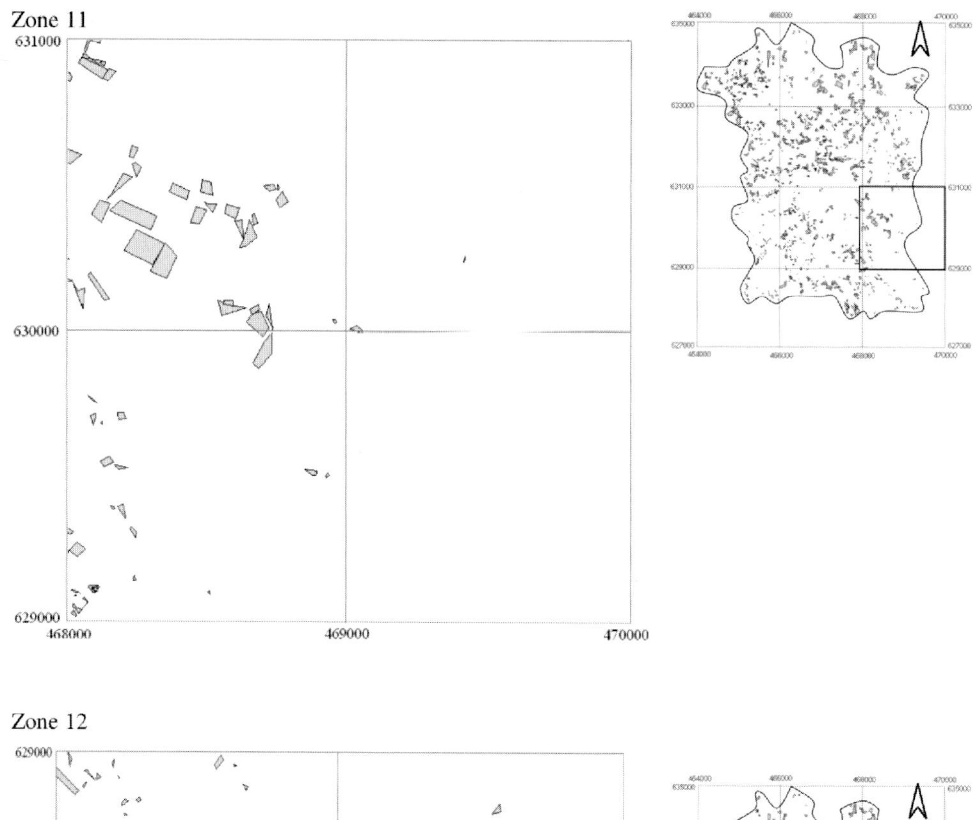

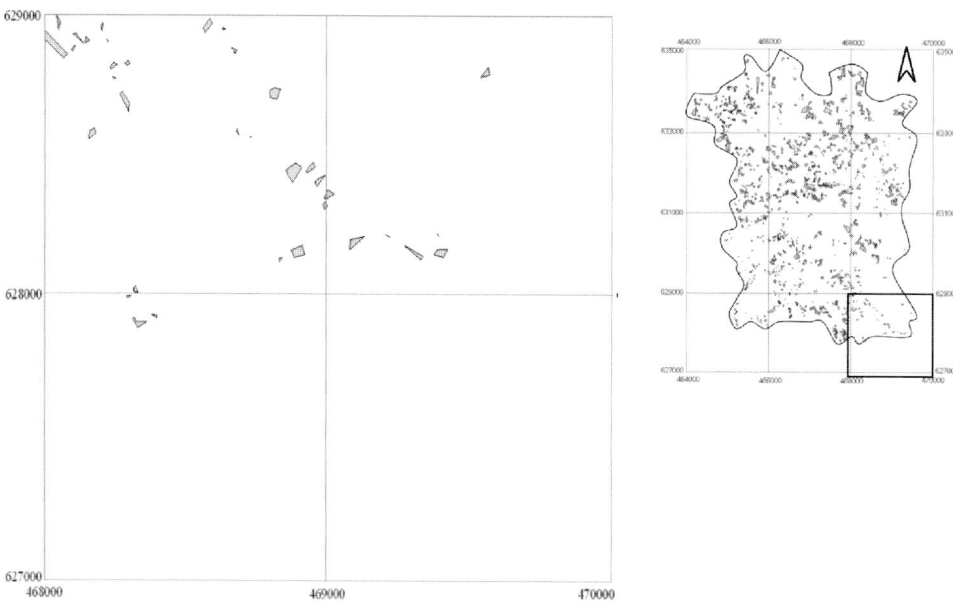

Figure 6.1.15: Spatial structure of maize fields in Zones 11 and 12. The relative positions of the frames are shown as inserts in the right top corner maps.

6.1.7 Field size distribution analysis

Seed and pollen dispersal is more likely to occur in major or dense cultivation regions. Therefore, it was of special interest to calculate the maize field sizes in relation to average maize field occurrence. Fig. 6.1.17 reveals high cultivation density characterized by smaller fields. The analysis shows a negative correlation emerging between the number of fields and field sizes, with field acreage mostly ranging between 100 -12,100 m^2 (0.01-1.20 ha) on average.

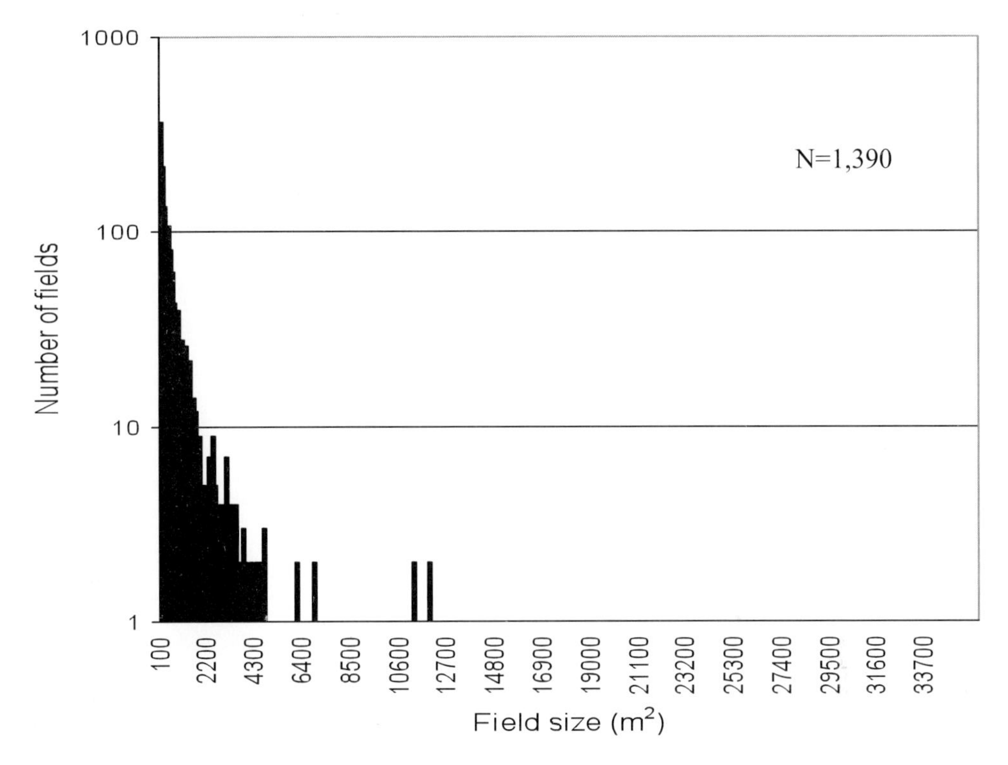

Figure 6.1.16: Acreage of maize fields in Accra West displayed as a histogram on a logarithmic scale indicating the nature of the agricultural structure basing on 1,390 maize fields in 25 km^2 areas.

6.1.8 Site characterization of finding places

Acreage of sites: Statistics of farm locations in relation to acreages are presented in Table 6.4. In terms of location, most sites where maize is grown location occur within home compounds constituting nearly 40% of all cases. Home compounds also represent the largest areas grown to maize, followed by traffic areas representing 30% of all locations with the second highest area covered with maize. Even though privately-owned open spaces have a lesser number of locations compared to construction sites, they constitute much larger area coverage. Low number of cultivated places is represented by industrial sites. In terms of field acreage, home compounds, traffic areas, construction sites, small business areas and marginal sites comprised the least acreages cultivated in some instances e.g. in a range of about 1-2 m^2. However, it was also revealed in some other instances that home compounds, traffic areas and construction sites for example had field acreage recorded going as high as 39,590m^2, 39,574m^2, and 27,082m^2 respectively (see Table 6.4).

Site characterisation of fields and ferals: Most of the cultivated plots found were within home compounds accounting for nearly 40% of all cases (Fig. 6.18). This was followed by locations around traffic areas, and construction sites estimated to about 29% and 11% respectively. Least occurrence of farms was found in industrial sites amounting to about 1%. Fig. 6.19 provides some examples of the situation. Majority of ferals (Fig. 6.21) were also found within home compounds representing about 38% of the cases. Then traffic areas and marginal sites each estimated to about 22%. Least locations are to be found in privately owned lands estimated to about 4%. No feral locations were found in wetlands as well as in small and industrial business areas (see Fig. 6.20).

Table 6.4: Maize field acreage within Accra West in 2006.

Site Description	Total No. of locations	Total field acreage (m^2)	Average field acreage (m^2)	Median field acreage (m^2)	Minimum field acreage (m^2)	Maximum field acreage (m^2)
Home compounds	550	408372.0	742.5	260.0	1.0	39590.0
Traffic areas	398	290527.0	730.0	272.5	2.0	39574.0
Privately-owned open places	59	179172.0	3036.8	1295.0	36.0	35132.0
Construction sites	147	171624.0	1167.5	318.0	1.0	27082.0
Public open spaces	96	69151.0	720.3	288.5	5.0	6087.0
Small business areas	55	48296.0	878.1	219.0	2.0	17668.0
Wetland areas	35	20555.0	587.3	227.0	10.0	5892.0
Industrial areas	13	8061.0	620.1	342.0	49.0	2251.0
Marginal sites	37	7507.0	202.9	91.0	2.0	1447.0
Total	1,390.0	1,203,265.0	8,685.0	3,313.0	108.0	174,723.0

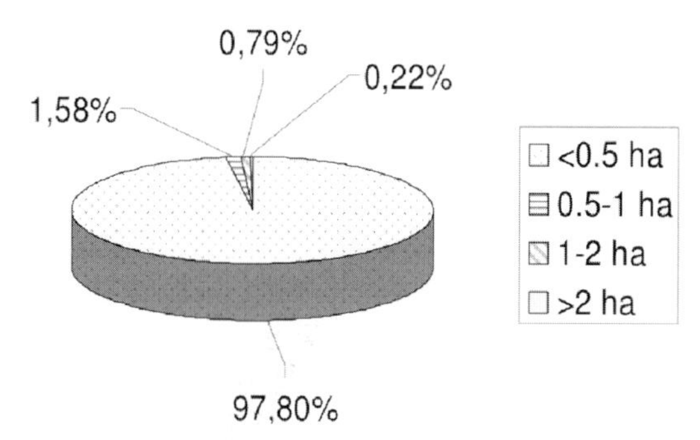

Figure 6.1.17: A pie chart showing the distribution range of farming acreage based on the 1,390 fields found. In all, about 97.5% of all locations are below half a hectare. Acreages above 2 hectare is least predominant calculated to be 0.22%. Farm sizes ranging between 0.5-1 hectare and 1-2 hectares are not very common estimated to be 1.58 and 0.79% respectively. This suggests that smaller fields are far larger in number and vice versa depicting that small-scale farming is typical in the region.

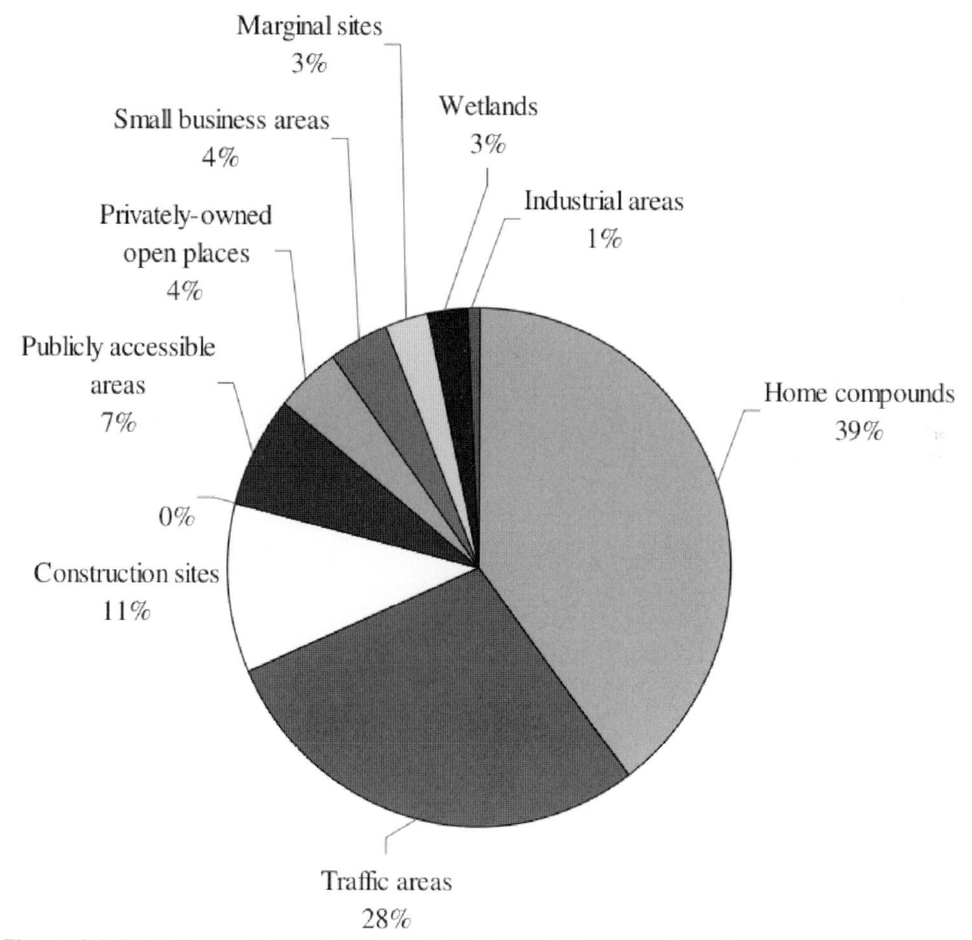

Figure 6.1.18: Percentage number of field locations within respective sites.

Figure 6.1.19: Examples of cultivated plots (a) Privately-owned open spaces; (b) Home compounds; (c) Publicly accessible areas; (d) Construction sites (e) Around wetlands; (f) Marginal sites within settlement areas.

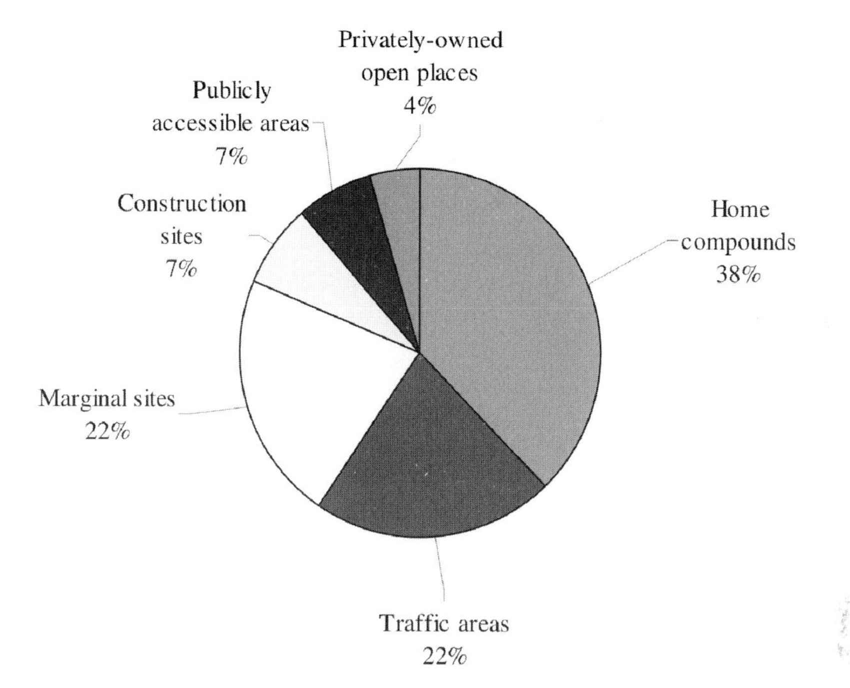

Figure 6.1.20: Percentage number of feral locations within respective sites; minimum number of feral locations is zero. Majority of feral was found within home compounds followed by traffic areas and marginal sites. Least locations were identified within privately-owned open spaces.

Figure 6.1.21: Examples of feral locations (a) Within a home compound: (b) Along a footpath; On a wall (e.g. Marginal site).

6.2 Crop Demographic Aspects

6.2.1 Stand frequencies and site characterization

Regression analysis of % farm locations and number of crop individuals (Fig. 6.2.1) yielded a strong negative correlation ($R^2 = 0.97$). Estimation for % feral locations in relation to individual single plants (Fig. 6.2.2) also gave a negative correlation ($R^2 = 0.80$). The data also reveals that most of the locations with less that 100 individuals were found within homes estimating to about 20% mostly gardens).

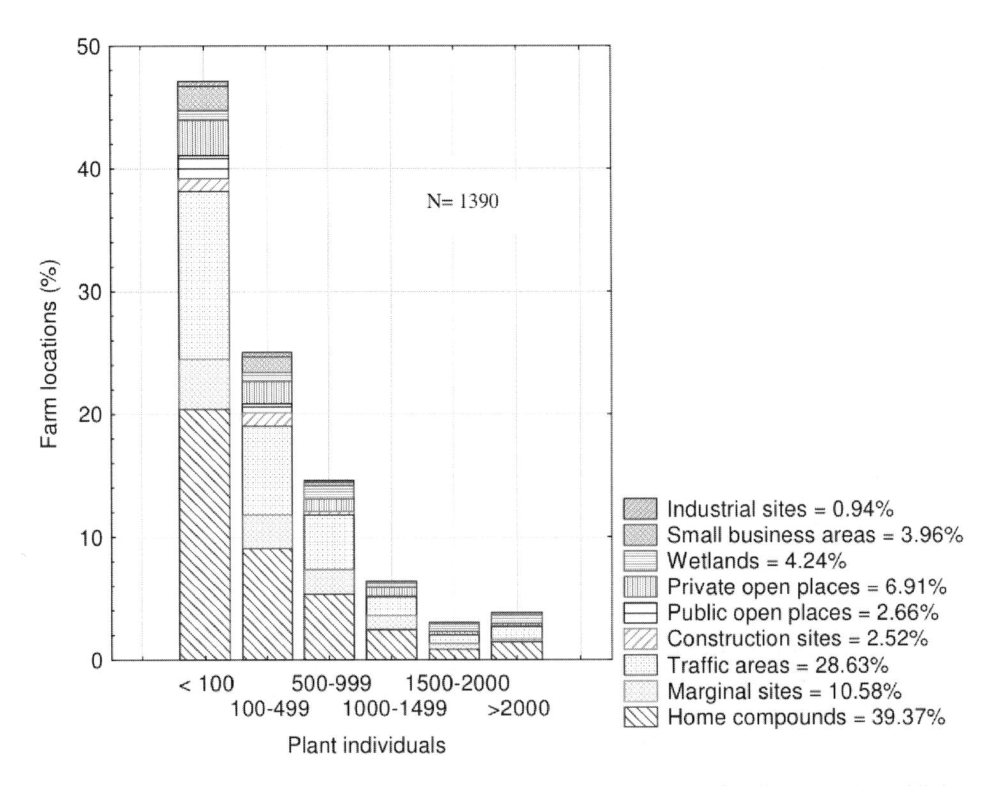

Figure 6.2.1: Crop demography in relation to site characteristics of cultivated plots. Highest population sizes are found in home compounds. Traffic areas constitute the second most important location with high population stands. Industrial locations have the least population of crops going up to a maximum less than 1000 individuals. The data show a wide variability in crop population sizes among the locations suggesting a limitation of space requirements in cultivation practices.

6.2.2 Farm neighbourhood frequencies under specified buffer conditions

The results indicate how many neighbours have to be taken into account on a specific scale (Fig. 6.2.4). Farm representativity on a level of increasing sampling points (or distances) is addressed. This is seen in a context of assessing co-existence implications of GM and non-GM crop cultivation within small-scale farming. The extent of monitoring for a GMO depends on the level of environmental exposure and needs to be considered on a case-by-case basis. E.g. exposure during import and processing is very limited compared to during cultivation (Lecoq et al., 2007).

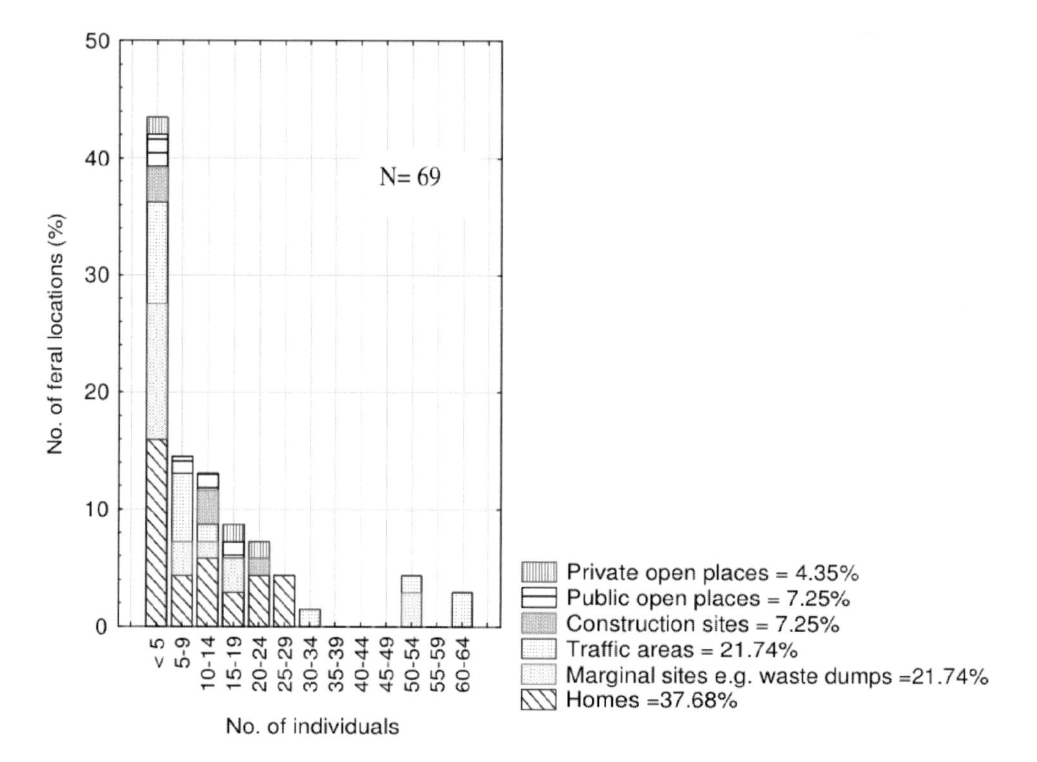

Figure 6.2.2: Feral demography in relation to their site characteristics. Most of the feral locations occur within home compounds but with the least population size of less than 5 individuals. Traffic and marginal areas are second most important places of feral occurrence. However traffic areas tend to have higher number of feral individuals than marginal sites suggesting higher possibility of seed losses during transport of seeds or harvests. Privately owned open spaces appear to have the least feral occurrence and population size. No feral stands were found in industrial, wetland, and in small business areas.

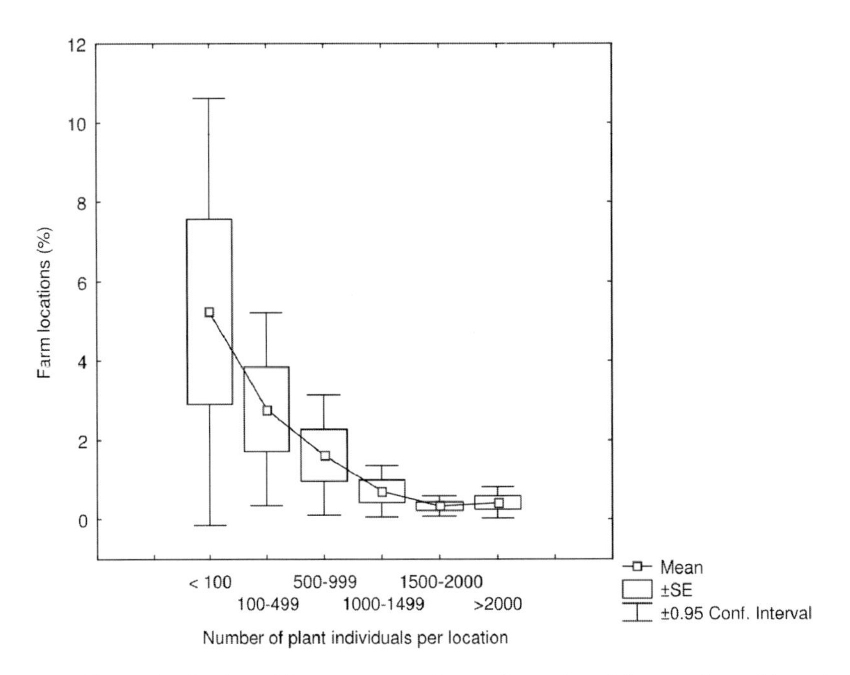

Figure 6.2.3: Mean distribution of farm crops at the various sites. The number of farm locations in relation to number of crop individuals shows a hyperbolic function. This indicates that larger crop populations occur within relatively few locations compared to smaller crop populations. For example, about 3%, 1.8%, and 0.9% of all locations have 100-499, 500-999 and 1000-1499 crop stands respectively.

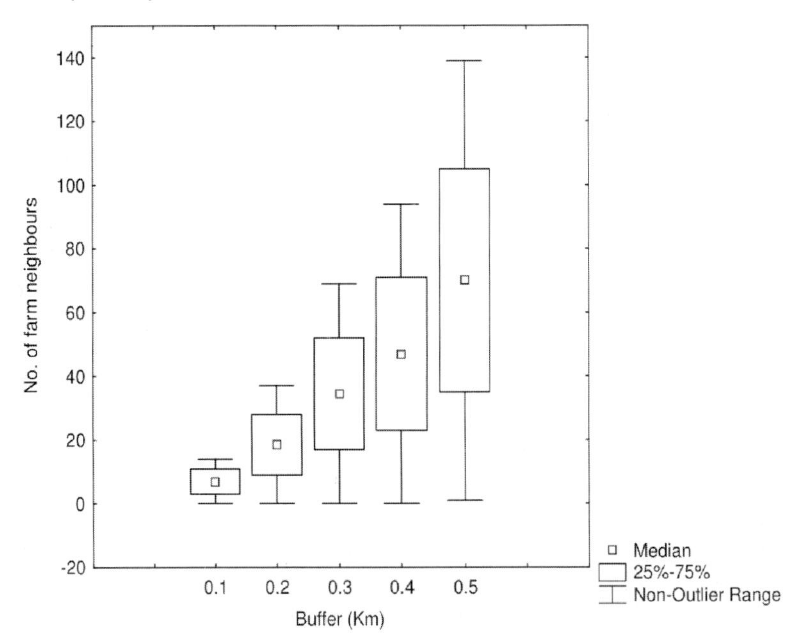

Figure 6.2.4: Number of farm neighbours within certain distance ranges (assumed here to be buffer). The data indicate a linear relation with an average number of 5, 19, 35, 48, and 70 farm neighbours occurring in the respective distances of 0.1, 0.2, 0.3, 0.4, and 0.5 Kilometres.

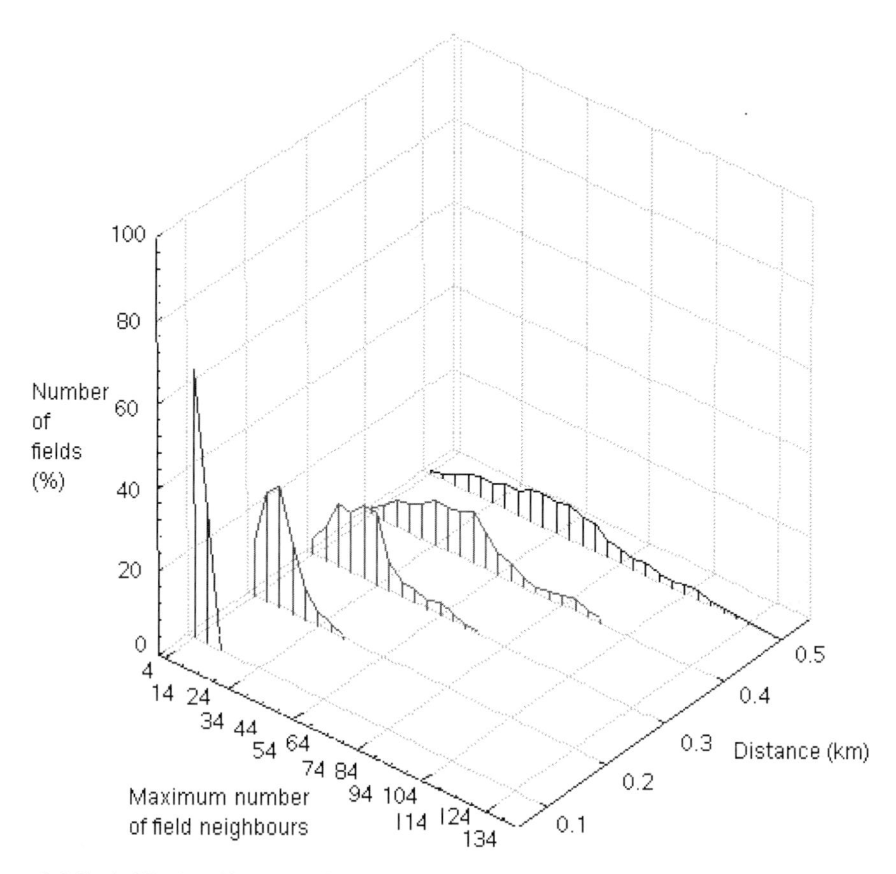

Figure 6.2.5: A 3D classification of farm neighbours to assess efficiency of isolation distance management based on the spatial structure. A relation of the number of fields to field interactions within certain distance ranges provides a useful way towards setting up a distance management regime for biosafety. The data show a higher number of interacting partners within larger distance ranges (buffers) when compared to shorter distances. For example, at 0.5 Km, about 17% of fields have 54 field neighbours while at a distance of 0.4 Km, 0.3 Km, 0.2 Km, and 0.1 Km, 17% of fields have 20, 14, 4, and 1 interacting neighbour(s). In the context of agricultural planning, regions with higher cropping densities would require to discuss among farmers for example delaying planting times or use of same varieties if farm cluster conditions are expected within a particular settlement area.

6.2.3 Crop phenological status and feral development data

Flowering rate: In Table 6.5 results of the flowering states of maize farms are shown in a 7-day (weekly) period. "Flowering rate" indicates the percentage of plants that simultaneously developed tassels and was calculated by averaging the number of flowering fields and weighting in % each type by its frequency. This could address the implied percentage field-to-field cross-pollination impacts within respective periods with impacts depending on the contribution effects of environmental conditions e.g. wind and rainfall factors. Peak flowering periods are revealed in June and July.

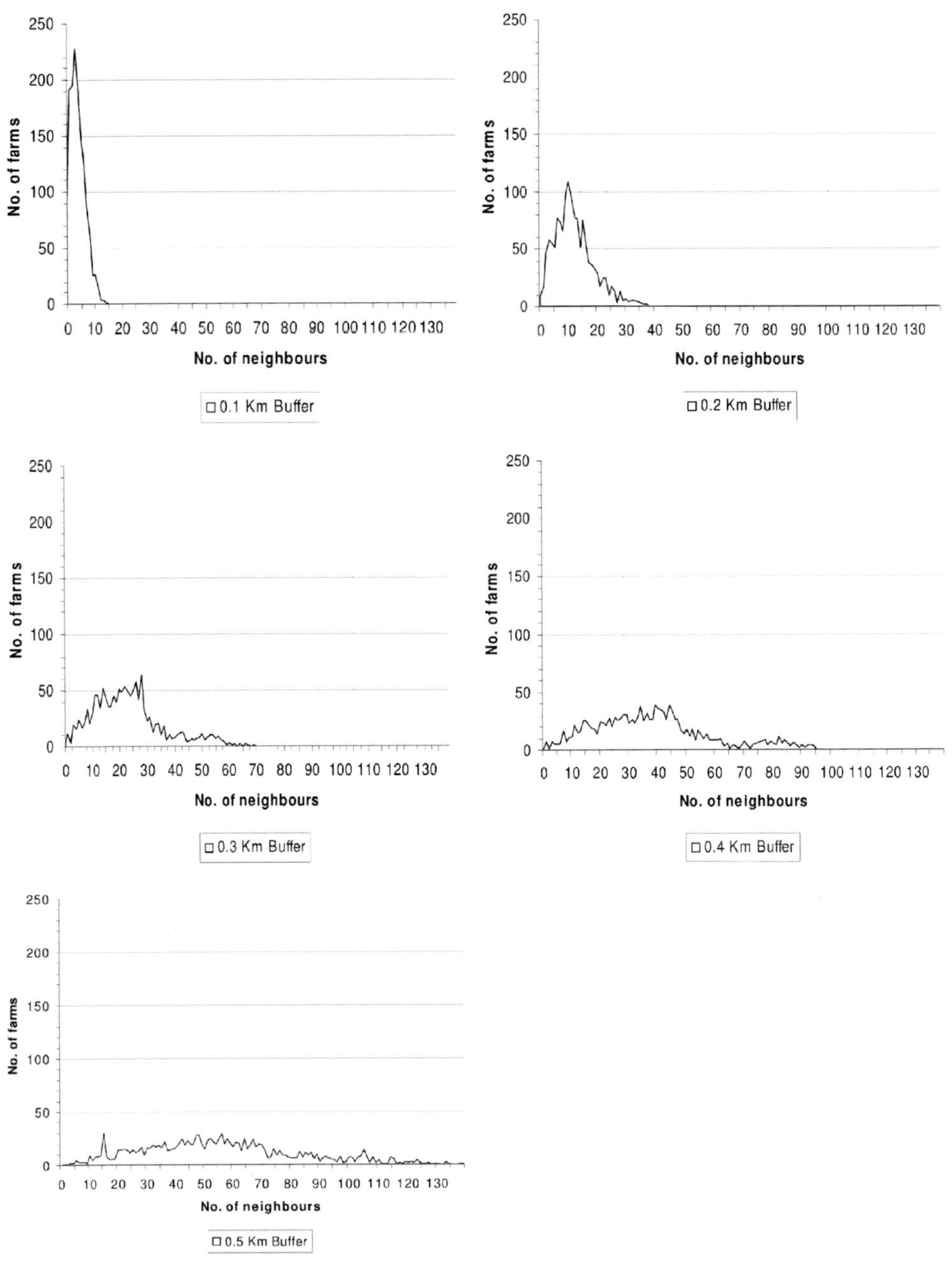

Figure 6.2.6: Number of farm neighbours of cases in 0.1, 0.2, 0.3, 0.4 and 0.5 km distances (N= 1,390 farms). Setting of barrier conditions is essential for estimating the number of neighbours, occurrence frequencies and hybridization probabilities across larger farming areas

In Fig. 6.2.6, a functional relationship estimating how many neighbours certain cluster of farms have within the selected distances of 0.1 km, 0.2 km, 0.3 km, 0.4 km, and 0.5 km is shown. For clarity, the information in Fig. 6.2.5 is repeated in a different form of display.

Table 6.5: Phenological characters of crop plants determined on weekly basis.

Sampling months (2006)	Pheno-logical states	Number of farms with concurrent phenological states					Onset of flowering rate (%)
		Week 1	Week 2	Week 3	Week 4	Locations found/ month	
May	1	13	37	6	-	56	27.14%
	2	55	45	34	-	134	
	3-4*	37	22	17	-	76	
	5-7	7	4	3	-	14	
	8+	0	0	0	-	0	
June	1	3	2	-	-	5	64.56
	2	73	80	-	-	153	
	3-4*	47	23	-	-	70	
	5-7	6	3	-	-	9	
	8+	0	0	-	-	0	
July	1	1	0	0	-	1	70.26
	2	20	18	5	-	43	
	3-4*	57	82	24	-	163	
	5-7	14	7	3	-	24	
	8+	0	1	0	-	1	
August	1	3	3	0	0	6	22.48
	2	19	16	8	8	51	
	3-4*	61	28	35	14	138	
	5-7	100	67	168	53	388	
	8+	0	3	16	10	31	
September	1	0	-	-	-	0	7.41
	2	0	-	-	-	0	
	3-4*	2	-	-	-	2	
	5-7	20	-	-	-	20	
	8+	5	-	-	-	5	
Total No. of locations =						1390	

Note: 1= Seedling (only embryonic leaves/ cotyledons); 2 = Beginning of longitudinal growth; 3-4* = First appearance of tassels (Onset of flowering period); 5-7 = Mature unripe (green) fruits/ cobs; 8+= Ripe (yellowing) cobs (completely withered silks); (-) denote non-sampling periods.

In Table 6.6 below, it can be estimated the % number of feral plants flowering simultaneously with crops. The data reveals considerable overlapping of flowering times between feral and farm crops. However, feral flowering rate in May is higher (41.03%) compared to those recorded for farms (27.14%). This suggests 2 things, the possibility of earlier emergence of volunteers from previous crop fields. Post-harvest crop management and handling of early sowing activities may be factors that determine their earlier presence and higher flowering rates. Peak flowering period of feral populations is recorded in July, and coincides with crop counterparts.

Table 6.6: Phenological characters of feral plants determined on weekly basis.

Sampling months (2006)	Absolute number of feral stands with concurrent phenological states						Onset of flowering rate (%)
	Pheno-logical states	Week 1	Week 2	Week 3	Week 4	Total No. of locations found/ month	
May	1	2	0	0	-	2	41.03%
	2	14	0	1	-	15	
	3-4*	3	13	0	-	16	
	5-7	0	5	0	-	5	
	8+	0	1	0	-	1	
June	1	0	0	-	-	0	22.22%
	2	6	1	-	-	7	
	3-4*	1	1	-	-	2	
	5-7	0	0	-	-	0	
	8+	0	0	-	-	0	
July	1	0	0	X	-	0	80.00%
	2	1	1	X	-	2	
	3-4*	4	4	X	-	8	
	5-7	0	0	X	-	0	
	8+	0	0	X	-	0	
August	1	0	0	X	X	0	45.45
	2	2	0	X	X	2	
	3-4*	3	2	X	X	5	
	5-7	1	3	X	X	4	
	8+	0	0	X	X	0	
Total No. of locations =						69	

Note: 1= Seedling (only embryonic leaves/ cotyledons); 2 = Beginning of longitudinal growth; 3-4* = First appearance of tassels (Onset of flowering period); 5-7 = Mature unripe (green) fruits/ cobs; 8+= Ripe (yellowing) cobs (completely withered silks); (-) denote non-sampling periods; and (x) imply no feral was found.

Flowering duration: The duration of flowering was measured to estimate average days of pollen exposure (viability) for different cultivars. The data (Fig. 6.2.7) suggests earlier and shorter pollen flowering duration ranging between 8-10 days (9 days. ±0.6 SD) observed for commercial varieties. Local hybrids showed slightly longer duration ranging between 9-11 days (10 days on average, ±0.8SD). Mixed varieties of commercial and local hybrids suggest a longer flowering period of 9-13 days (11 days on average, ±1.3SD).

Feral distances to crop in flower: There were 69 feral compared to 1390 field locations amounting to a ratio of 1:20 on the landscape. An understanding of the extent to which transgene movement from GM ferals could potentially contribute to the overall maize crop gene pool is relevant for regulatory management. Therefore, pollen transfer distances from feral locations to nearest crop in flower are analyzed in Fig. 6.2.8 below.

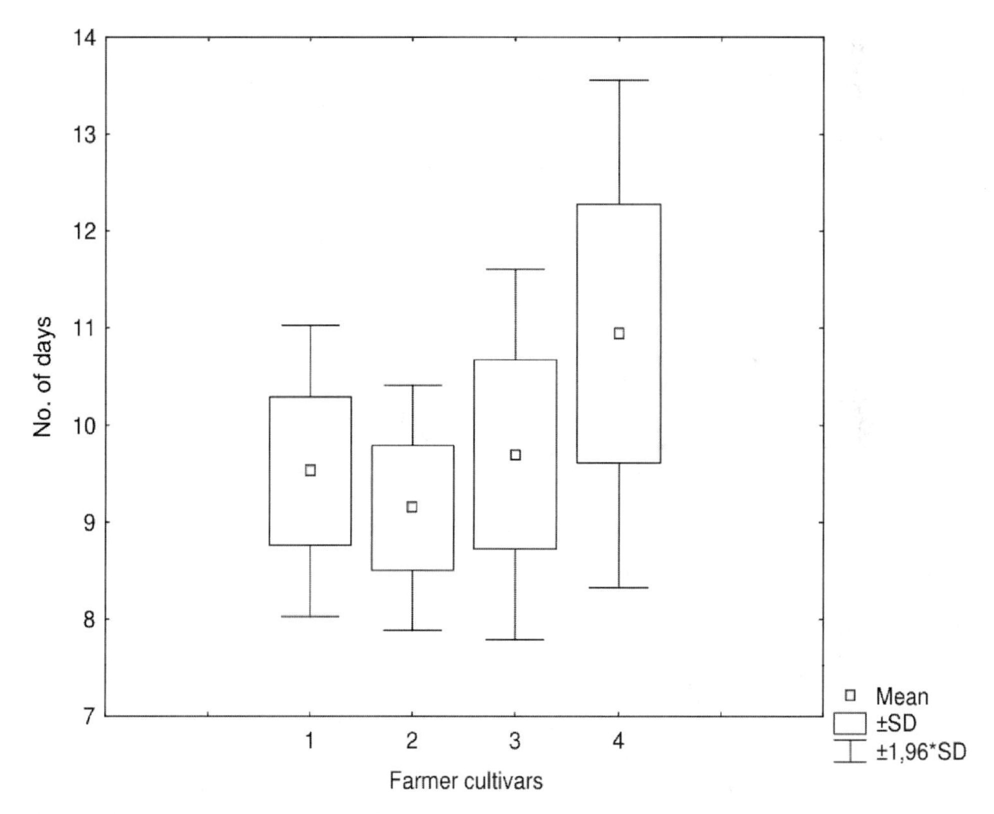

Figure 6.2.7: Flowering duration of 4 farmer cultivars using temporal difference from onset to shed of pollen. OPV seed varieties (1), Commercial seeds mostly Obaatanpa (2); 'Other improved variety' (name unknown to farmer); Mixed hybrid of local and commercial seeds (4). The results show the averages of randomly pooled data from 6 fields each for local and commercial cultivars.

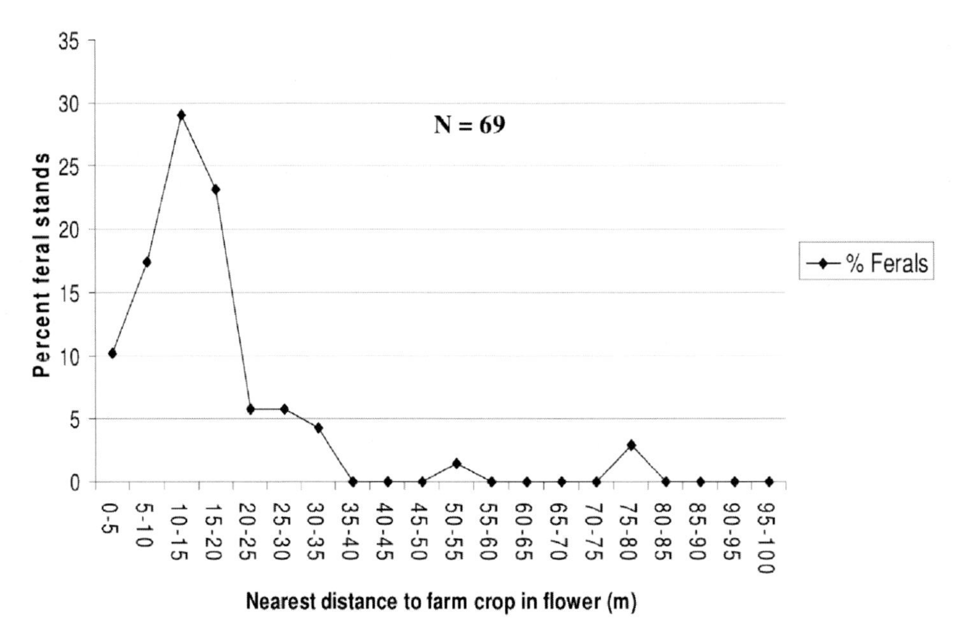

Figure 6.2.8: Estimation of cross-pollination potential between cultivated crops and feral locations. The data shows that 10%, 23%, and 28% of all feral locations occur within 5 m, 10 m, and 15 m respectively from flowering farm crops. Longest distance to a farm crop in flower was recorded at a distance range of 75-80 m. This poses implications for gene flow between crop plants and feral stands in the wild.

Figure 6.2.9: Feral stand in groundnut field few meters (ca. 75 m) from farm crops

6.2.4 Summary of results of geographical and crop demographic data

The implications for gene flow using cropping distances, flowering and population demographic factors of crops are summarized in Table 6.7.

Table 6.7: Gene flow conditions basing on geographical and crop demography.

Factor	Descriptor	Result
(1) Flowering rate	Feral flowering simultaneously with crop (%)	47.2%
	Feral plants as % of total maize crops flowering	15.5%
(2) Wild feral demography	% of feral populations within distance to nearest crop in flower 5 m 10 m 15 m 35 m 80 m	 10% 17% 29% 4% 3%
	Likelihood of gene flow from crops to feral (89.9% of feral stands within 85 m from farms)	Very likely
	Feral neighbours within certain distances (random selection), mean (and range) 0-100 m 300-400 m 1500-1600 m 2000-2100 m 3900-4000 m	 1 (0-3) 2 (0-6) 2 (0-7) 3 (0-11) 16 (8-26)
(3) Feral site features	Habitats (%) of feral within: Homes Traffic areas Marginal sites (e.g. soil dump, Farmyard, garden) Construction sites Public open spaces Privately-owned open spaces	 37.7% 21.7% 21.7% 7.3% 7.3% 4.4%
	Possible origins local recruitment (e.g. domestic activities, shatter from farms) vehicular transport	 37% 51%
(4) Feral distances and neighbourhood relations	% feral stands lying in nearest distance range of 5-100m	46.4%
	% feral with shortest nearest distance of 16m	5.8%
	% feral with the longest nearest distance of 1,548m	1.45%
	Feral stand density (km^{-2}) mean	3
(5) Farm density and number of farms	Farm stands density (km^{-2}) mean	56
	Total number of farm plots flowering (km^{-2}) all populations	17.8%
(6) Farm site features	Habitats (%) of farms on Homes (e.g. as gardens) traffic areas Construction sites Public open spaces Marginal sites (e.g. soil dump, Farmyard, garden) Wetlands	 39.6% 28.6% 10.6% 6.9% 2.7% 2.5%

Factor	Descriptor	Result
(7) Farm distances and neighbour-hood relations	% farm stands lying in nearest distance range to each other of 5-100 m	98%
	% farm plots with shortest nearest distance of 5 m to each other	4.32%
	Farm neighbours within certain distance ranges. Mean (and range) 0-100 m 300-400 m 1500-1600 m 2000-2100 m 3900-4000 m	 4 (0-14) 15 (0-41) 53 (22-90) 70 (30-137) 151 (93-226)
	% farm plots with the longest nearest distance of 459 m	0.072%
(8) Farm acreage	% of farms with acreage of: < 0.5 ha 0.5-1 ha 1-2 ha > 2 ha	 97.48% 1.58% 0.79% 0.22%

6.3 Ecological Modeling of Cross-Pollination Probabilities

6.3.1 Simulation of regional cross-pollination between GM and conventional small-scale maize cropping systems

(a) Model Scenario 1- GM Seeds planted obtained under farmer exchange conditions: Fig. 6.3.1 depicts the locations and sizes of conventional maize farming practices in Accra. The dark locations are the assumed GM fields grown among conventional fields (shown grey). Dark locations generally typify areas where seeds planted had been previously obtained from other farmers, as gifts, borrowed, exchanged or simply covert acquisitions as revealed from the socioeconomic data. The locations do not represent all locations (n=1,390) with exchanged seeds – but only those identified from the farmer interviews (n=201), of which 42 stated to have received seeds through exchange with other farmers. As a scenario assumption, these were calculated as GM. In Figs. 6.3.2, model results indicating the spatial nature of emerging cross-pollination rates between the GM fields and conventional crops are shown. The dispersal function for both 'modern hybrids' (MH) and (b) 'open pollinated' (OPVs) have been applied for this case scenario.

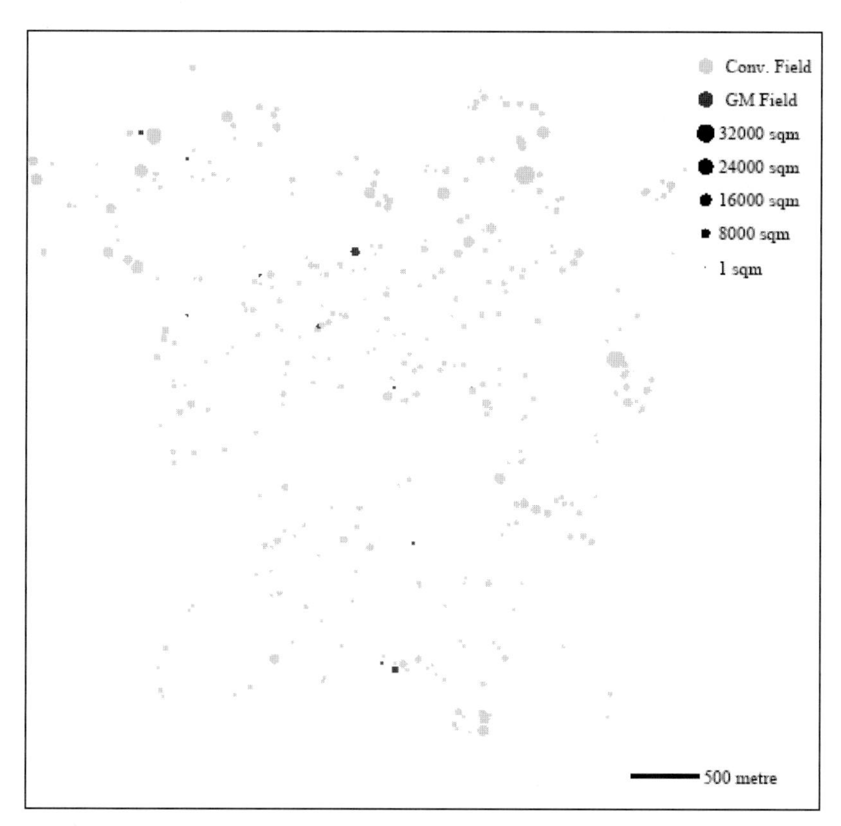

Figure 6.3.1: Spatial configuration of 42 GM fields among 1,348 conventional fields. This is an initial model map with stipulated criteria that GM fields were cultivated from farmer exchanged seeds. GM fields have a variable size range of up to 8000 m^2.

Examples of simulation of pollen hybridization rates (%) between 42 small-scale GM and 1,348 conventional fields is shown on next page.

Figure 6.3.2: Simulation of pollen hybridization rates (%) between small-scale GM and conventional fields presented on a logarithmic scale. The total number of GM fields planted in this scenario is 42, and considers that the GM seeds planted were obtained under farmer exchange conditions. This condition is applied hypothetically to depict potential contamination rates basing on the seed exchange practices among farmers as obtained from the socioeconomic data (a) Presents the result of the applied dispersal function of 'modern hybrids' (MH) and (b) applied the 'open pollinated' (OPVs) dispersal function. Average presence of GM input in conventional harvests is higher for OPVs compared to MHs amounting to 1.4 and 0.9 respectively. This means that 0.5% of the fields are sufficient to contaminate the whole area up to the EU labeling threshold of 0.9% above of which a crop cannot be sold as conventional harvest.

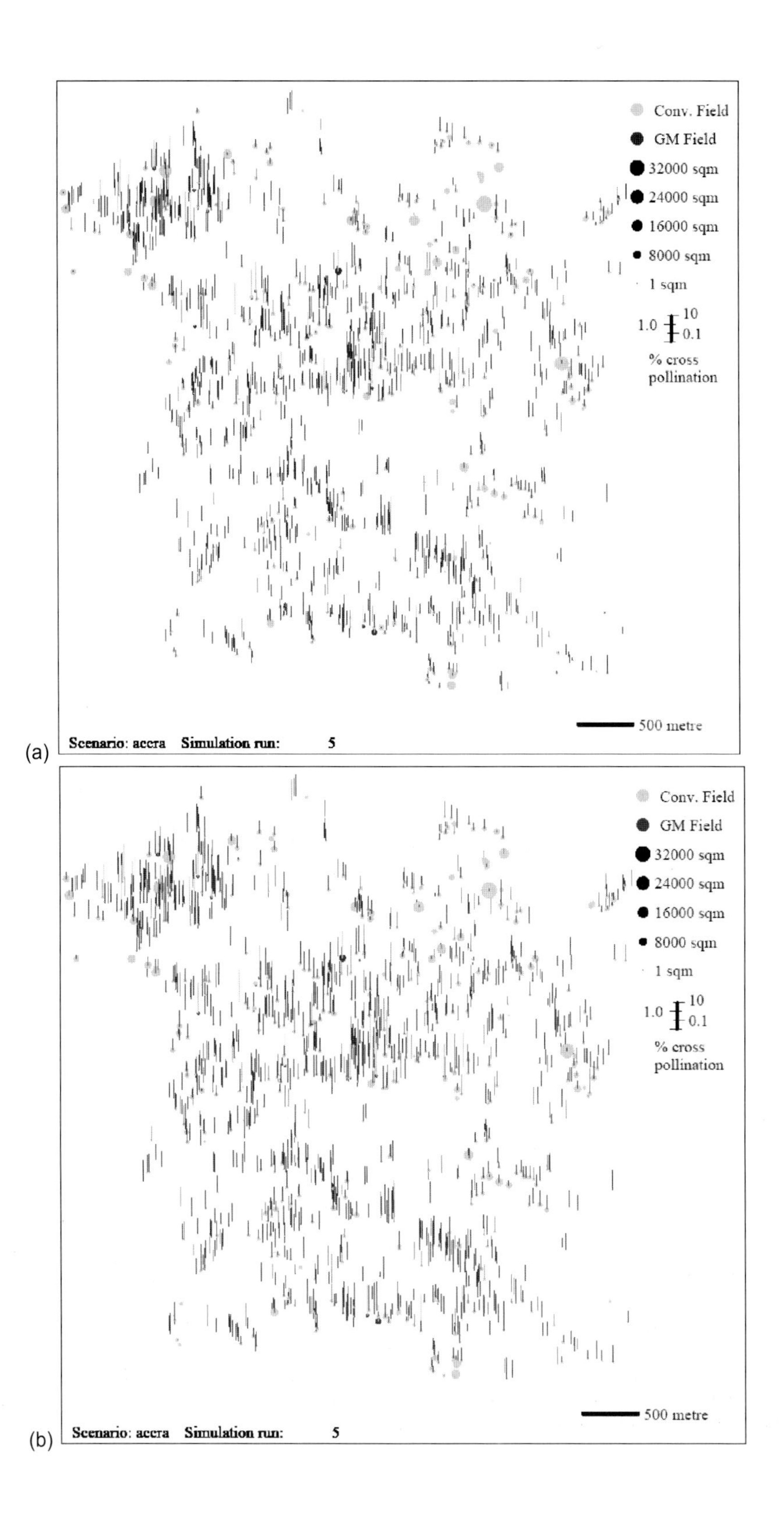
Conv. Field
GM Field
32000 sqm
24000 sqm
16000 sqm
8000 sqm
1 sqm
1.0
10
0.1
% cross pollination
500 metre
Scenario: accra Simulation run: 5
(a)
Conv. Field
GM Field
32000 sqm
24000 sqm
16000 sqm
8000 sqm
1 sqm
1.0
10
0.1
% cross pollination
500 metre
Scenario: accra Simulation run: 5
(b)

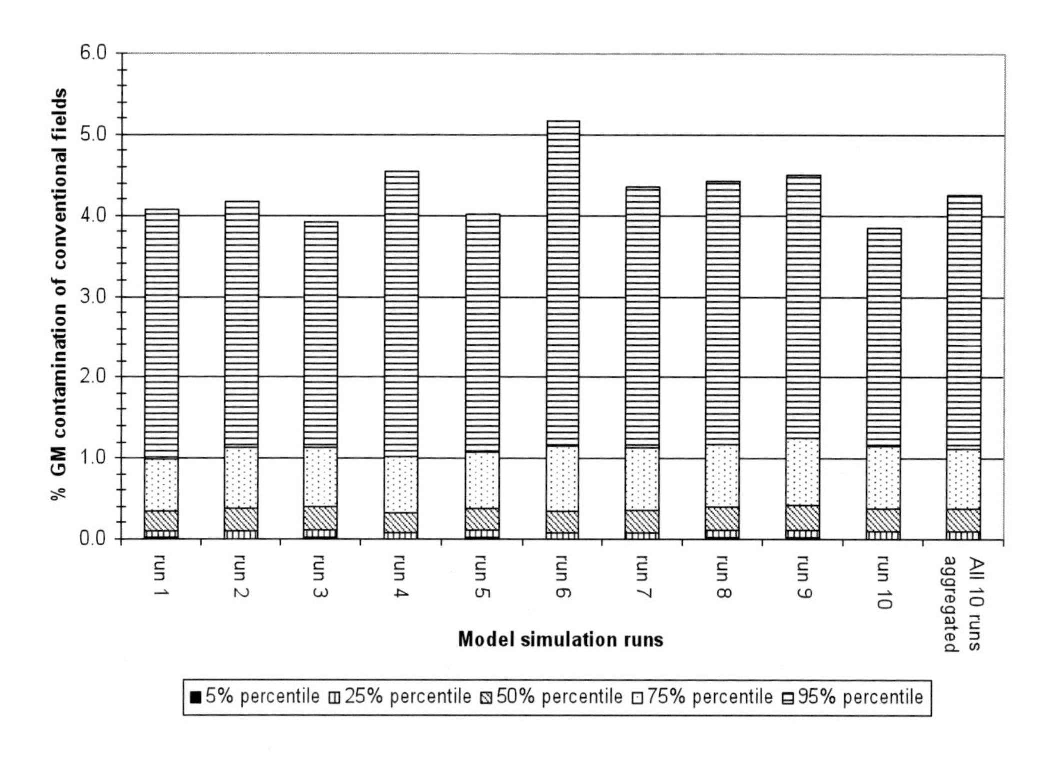

Figure 6.3.3: Quantitative estimation in percentiles of the mean GM pollination rates of conventional fields based on the farmer seed exchange criteria for the 10 repeated model runs. The variability range of GM cross-pollination for all 10 runs aggregated is also presented. The derived variability as shown is a result of the Monte Carlo simulations that considered each single field as a source and as a receptor in cross-pollination. Highest variability of cross-pollination is most pronounced in the range of 1.0%-5.3% for all cases (i.e. occurring within the uppermost percentile of 95%). This indicates an overall higher GM cross-pollination occurring in this range for all conventional fields. Most probable (stable) estimate of cross-pollination under the set of criteria may be described at the 50% percentile (i.e. at the median) in the value range from 0.1%-0.3% cross-pollination.

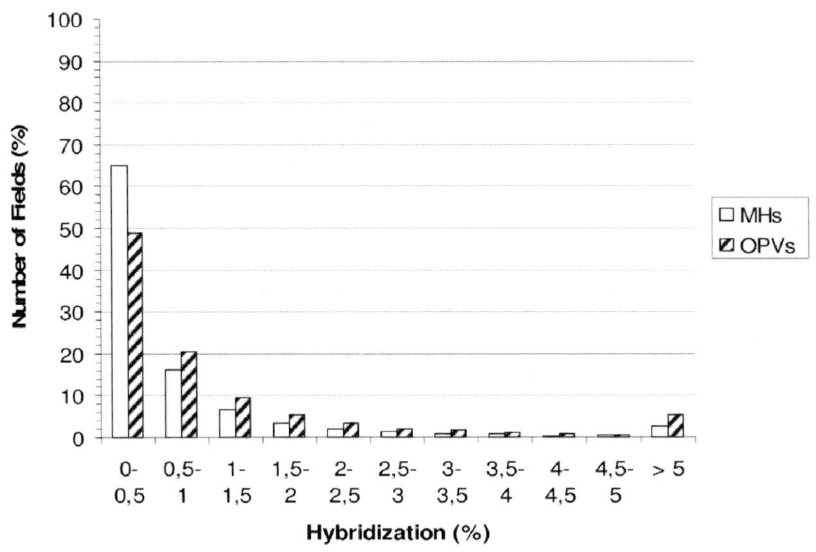

Figure 6.3.4: Model result of GM presence in conventional fields under seed exchange among farmers involving modern hybrids (MHs) or open-pollinated varieties (OPVs). The data depicts that a lot of conventional fields of about 65% have nearly no GM input for MHS, and 49% for OPV scenario. However, 16% and 20% of the fields have GM input in the range of 0.5-1.0 for MHs and OPVs respectively. The data shows a generally higher GM input for the OPVs as compared to the MHs. Hybridization rates extending beyond 5% is recorded in 4% and 7% of all fields for the MHS and OPVs respectively.

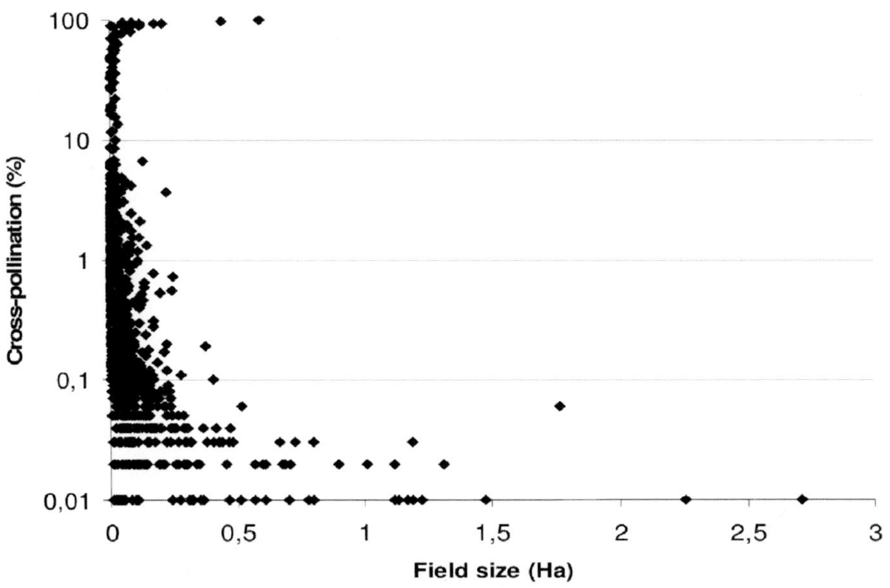

Figure 6.3.5: GM cross-pollination rates in 1,348 conventional fields in relation to acreage displayed on a logarithmic scale. This is based on the assumption that GM seeds were obtained through farmer exhanges. The relation is hyperbolic function revealing a gradual decrease in cross-pollination rates with increasing size of recipient field and vise- versa. Cross-pollination of nearly 100% is obtained for few fields in a size range of about a few square meters (e.g. ~1-5 m^2).

(b) Model Scenario 2: GM Seeds planted obtained from the food market: Fig. 6.3.3 depicts the locations and sizes of conventional maize farming (shown grey). The dark locations show locations where seeds planted had been previously obtained from the food market as revealed from the socioeconomic data, assumed to be GM fields to simulate emerging properties of cross-pollination rates among the GM and conventional fields presented in the map. As in the previous scenario, the responses of the farmers in the socioeconomic survey were used to locate the cultivation sites. 35 fields of 1,390 were assumed to be GM. This simulates the entry path that commercial harvests are used as seeds in the region under consideration.

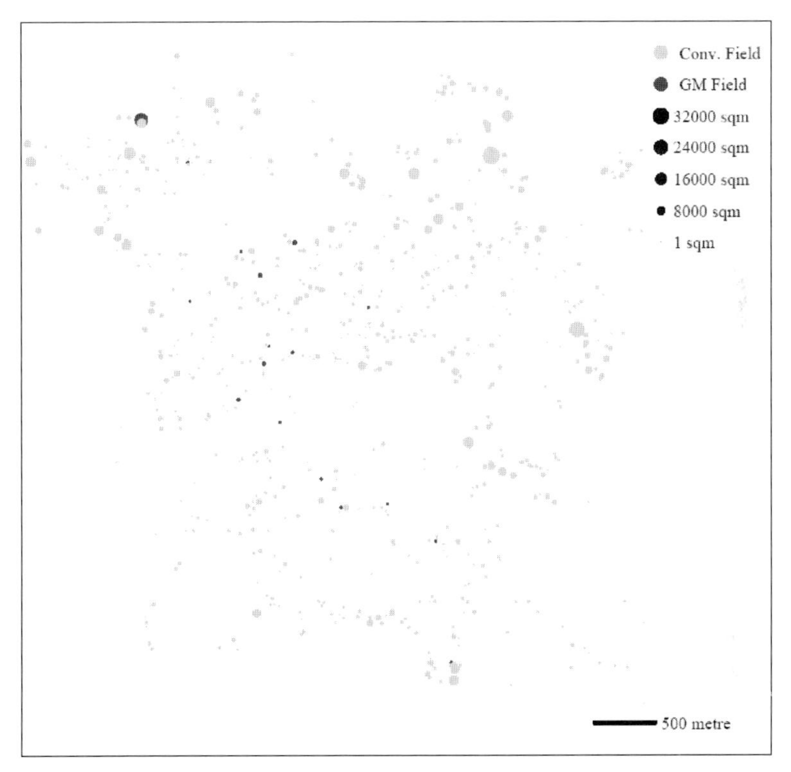

Figure 6.3.6: Initial model assumption stipulating the criteria that fields planted with GM utilised seeds obtained from the food market. This map scenario was developed with the help of the socioeconomic data. The total number of GM fields planted in this scenario is 35, and considers that the GM seeds planted were acquired from the food market.

Examples of simulation of pollen hybridization rates (%) between 35 small-scale GM and 1,355 conventional fields is shown on next page.

Figure 6.3.7: Simulation of pollen hybridization rates (%) between small-scale GM and conventional fields presented on a logarithmic scale. The total number of GM fields planted in this scenario is 35, and considers that the GM seeds planted were acquired from the food market. This is a hypothetical condition to depict potential contamination rates basing on the seed acquisition practices among farmers as obtained from the socioeconomic data. (a) Presents the result of the applied dispersal function of 'modern hybrids' and (b) applied the 'open pollinated' dispersal function.

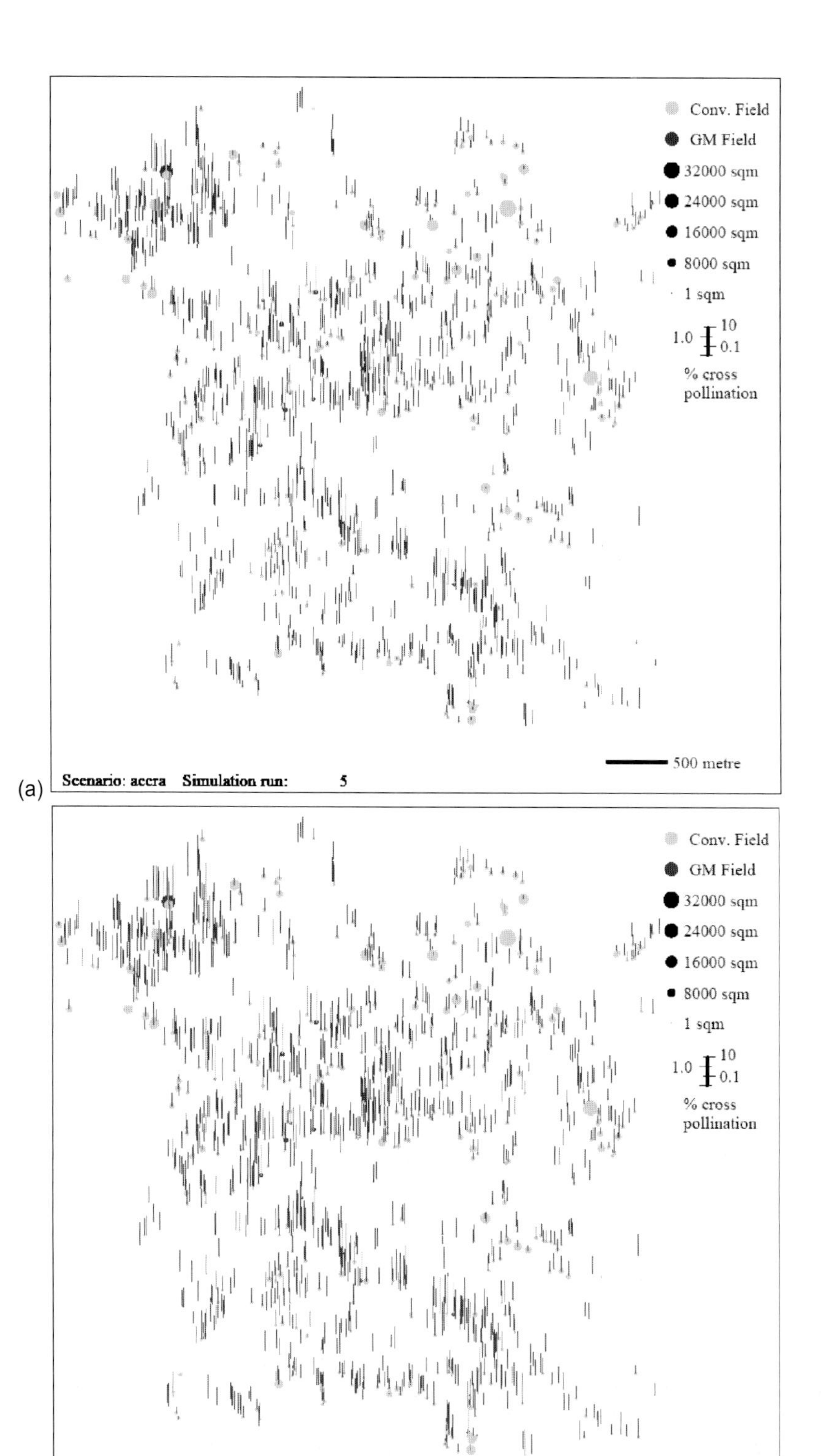
Conv. Field
GM Field
32000 sqm
24000 sqm
16000 sqm
8000 sqm
1 sqm
1.0
10
0.1
% cross pollination
500 metre
Scenario: accra Simulation run: 5
(a)
Conv. Field
GM Field
32000 sqm
24000 sqm
16000 sqm
8000 sqm
1 sqm
1.0
10
0.1
% cross pollination
500 metre
Scenario: accra Simulation run: 5
(b)

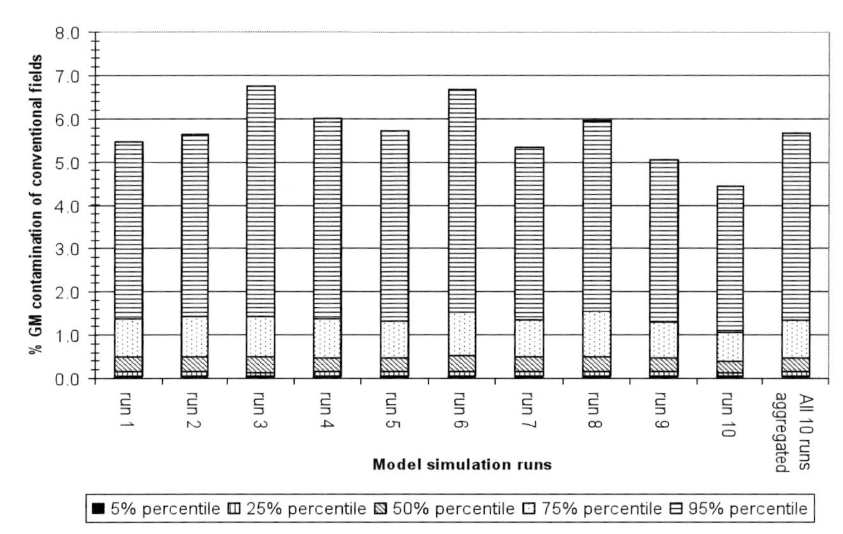

Figure 6.3.8: Quantitative estimation in percentiles of the mean GM pollination rates of conventional fields with the assumption that the GM sources were from food grain (market). Additionally, the variability range of GM cross-pollination for all 10 runs aggregated is shown. The derived variability is a result of the Monte Carlo simulations that considered each single field as a source and as a receptor in cross-pollination. Highest variability range of cross-pollination is most pronounce from 1.0%-6.7% for all the cases (i.e. occurring within the uppermost percentile of 95%). Thus, indicating an overall higher GM cross-pollination rate in this range among all conventional fields. Most probable (stable) estimate of cross-pollination under the set of criteria may be described at the 50% percentile (i.e. at the median) in the value range from 0.1%-0.4% cross-pollination.

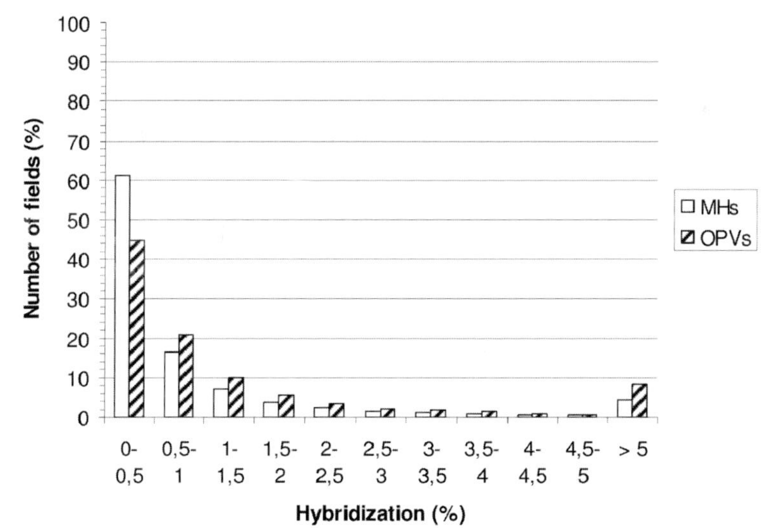

Figure 6.3.9: Model result of GM presence in conventional fields due to introductions from food market, assumed to be modern hybrids (MHs) or open-pollinated varieties (OPVs). The data indicates that a majority of conventional fields of about 61% have nearly no GM input for MHS, and 45% for OPV scenario. However, 16% and 22% of the fields have GM input in the range of 0.5-1.0 for MHs and OPVs respectively. Generally, higher GM inputs for the OPVs as compared to the MHs are observed. Hybridization rates extending beyond 5% is recorded in 6% and 8% of all fields for the MHS and OPVs respectively.

(c) Model Scenario 3: GM Seeds planted obtained from the seed market and extension services: Fig. 6.3.5 also shows the locations and sizes of conventional maize farming (shown grey in map). The dark places indicate locations where seeds planted had been bought from extension agents or from the seed market, as revealed from the 201 farmers interviewed in the socioeconomic survey. This seed acquistion information has been assumed to be GM fields to simulate emerging properties of cross-pollination rates among the GM and conventional fields presented in the map.

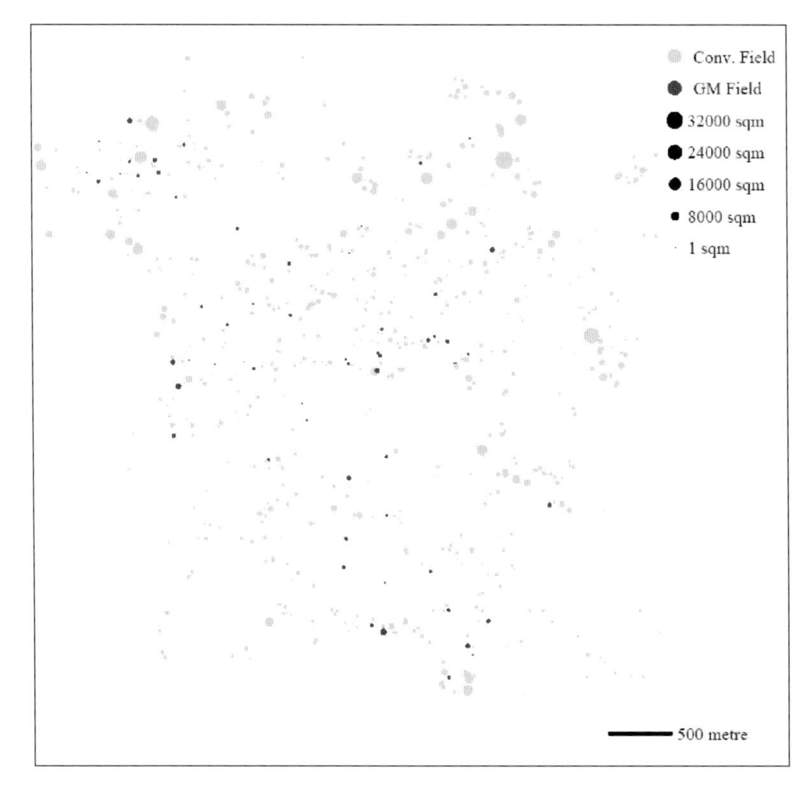

Figure 6.3.10: Initial model assumption stipulating the criteria that fields planted with GM utilised seeds obtained from the seed market and extension services. This map scenario was developed with the help of the socioeconomic data. The total number of GM fields planted in this scenario is 107 occurring among 1,283 fields and considers that the GM seeds planted were procured from the seed market and extension agents.

Examples of simulation of pollen hybridization rates (%) between 107 small-scale GM and 1,283 conventional fields is shown on next page.

Figure 6.3.11: Simulation of pollen hybridization rates (%) between small-scale GM and conventional fields presented on a logarithmic scale. The total number of GM fields planted in this scenario is 107, and considers that the GM seeds planted were procured from the seed market and extension agents. This is a hypothetical condition to depict potential contamination rates basing on seeds bought for planting from formal sources (e.g. shops) as obtained from the socioeconomic data. (a) Presents the result of the applied dispersal function of 'modern hybrids' and (b) used the 'open pollinated' dispersal function.

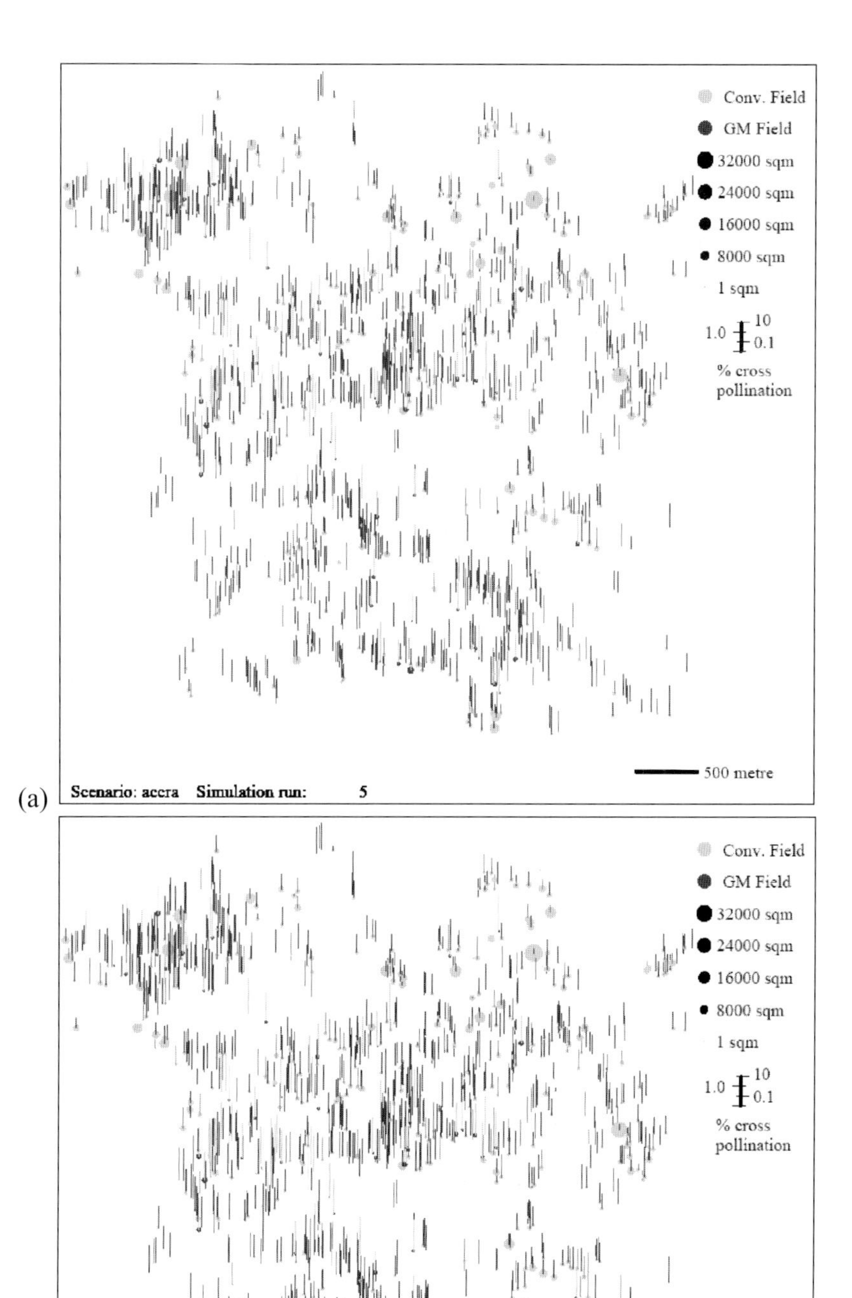
Conv. Field
GM Field
32000 sqm
24000 sqm
16000 sqm
8000 sqm
1 sqm
1.0
10
0.1
% cross pollination
500 metre
Scenario: accra Simulation run: 5
(a)
Conv. Field
GM Field
32000 sqm
24000 sqm
16000 sqm
8000 sqm
1 sqm
1.0
10
0.1
% cross pollination
500 metre
Scenario: accra Simulation run: 5
(b)

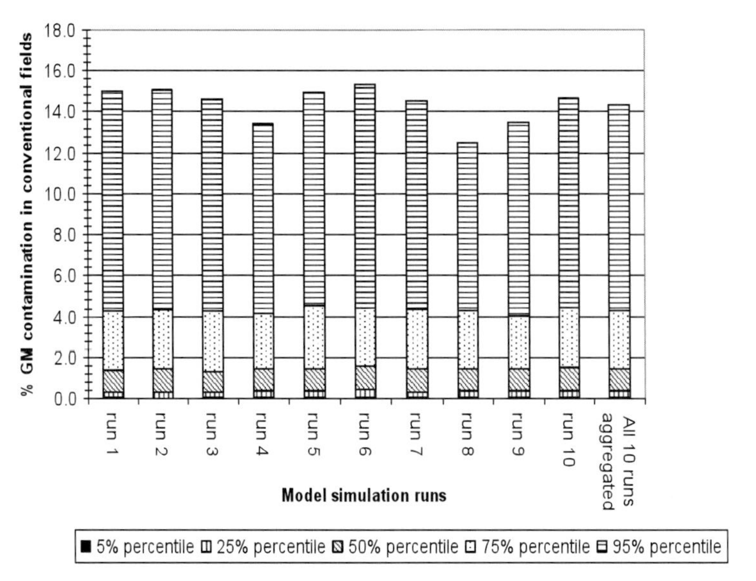

Figure 6.3.12: Quantitative estimation in percentiles of the mean GM pollination rates of conventional fields with the assumption that the GM sources were from the seed market and extension. The variability range of GM cross-pollination for all 10 runs aggregated is also presented. The derived variability is a result of the Monte Carlo simulations that considered each single field as a source and as a receptor in cross-pollination. Highest variability range of cross-pollination is most pronounced from 4.0%-14.7% for all the cases (i.e. occurring within the uppermost percentile of 95%). Thus, indicating an overall higher GM cross-pollination rates in this range among all conventional fields. Most probable (stable) estimate of cross-pollination under the set of criteria may be described at the 50% percentile (i.e. at the median) in the value range from 0.2%-0.7% cross-pollination.

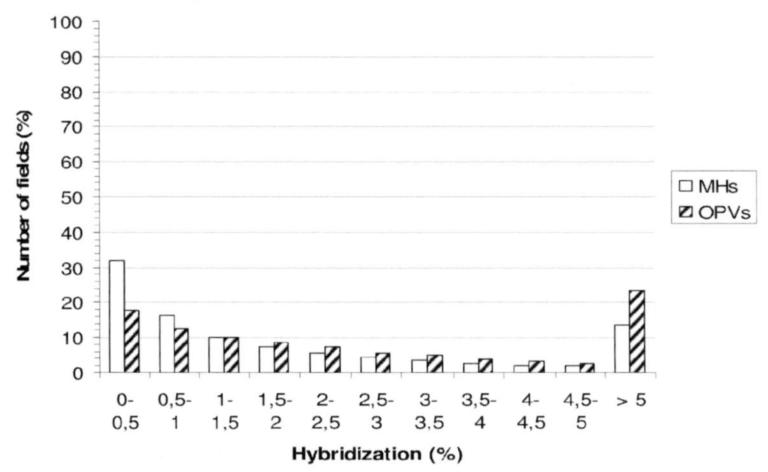

Figure 6.3.13: Model result of GM presence in conventional fields due to introduction from seed market and extension services assumed to be modern hybrids (MHs) or open-pollinated varieties (OPVs). The data indicates that a majority of conventional fields of about 32% have nearly no GM input for MHS, and 18% for OPV scenario. 17% and 14% of the fields have GM input in the range of 0.5-1.0 for MHs and OPVs respectively. Generally, higher GM inputs for the OPVs as compared to the MHs are observed. Hybridization rates extending beyond 5% is recorded in 14% and 24% of all field cases for the MHS and OPVs respectively.

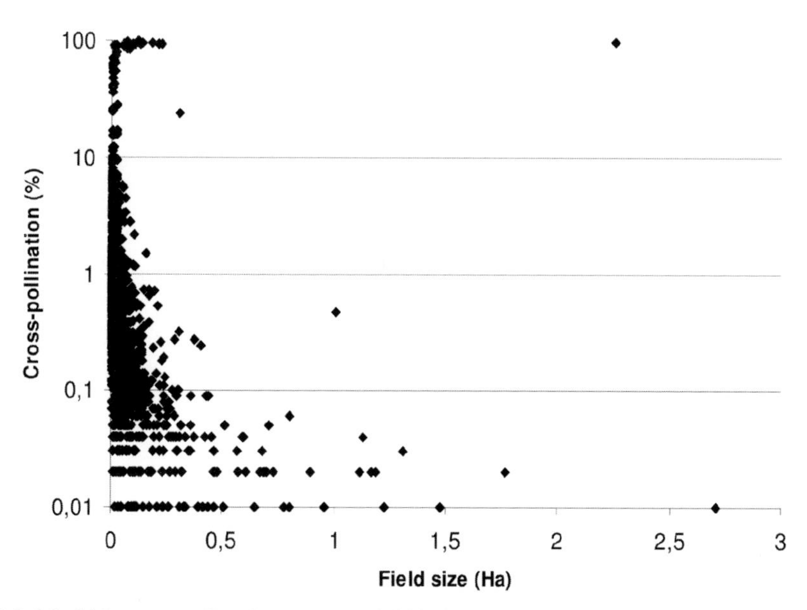

Figure 6.3.14: GM cross-pollination rates in 1,283 conventional fields in relation to acreage displayed on a logarithmic scale. This is based on the condition that GM seeds were obtained from the seed market and through extension services. The relation is hyperbolic function revealing a gradual decrease in cross-pollination rates with increasing size of recipient field and vise- versa. Cross-pollination of nearly 100% is obtained for few fields in a size range of about a few square meters (e.g. ~ 1-5 m^2).

(d) Model Scenario 4: Single GM field in the centre of the study area: Since the model uses each field as a pollen source and calculates the impact to all others, it is difficult to estimate the influence of a single field separately. To show the implied random influences and influence of size relations, we present here a simulation with only one field located in the centre of the investigated region.

Examples of simulation of pollen hybridization rates (%) between 1 small-scale GM and 1,389 conventional fields is shown on next page.

Figure 6.3.15: Model result stipulating the criteria of a single centre GM field among conventional fields numbering 1,389. (a) Shows the initial map with the centre field within the circle (b) Shows the results of cross-pollination rates of the fields. There are some conventional fields that do not receive any GM input even though close in proximity to the GM field. These points the random processes involved in pollen movement.

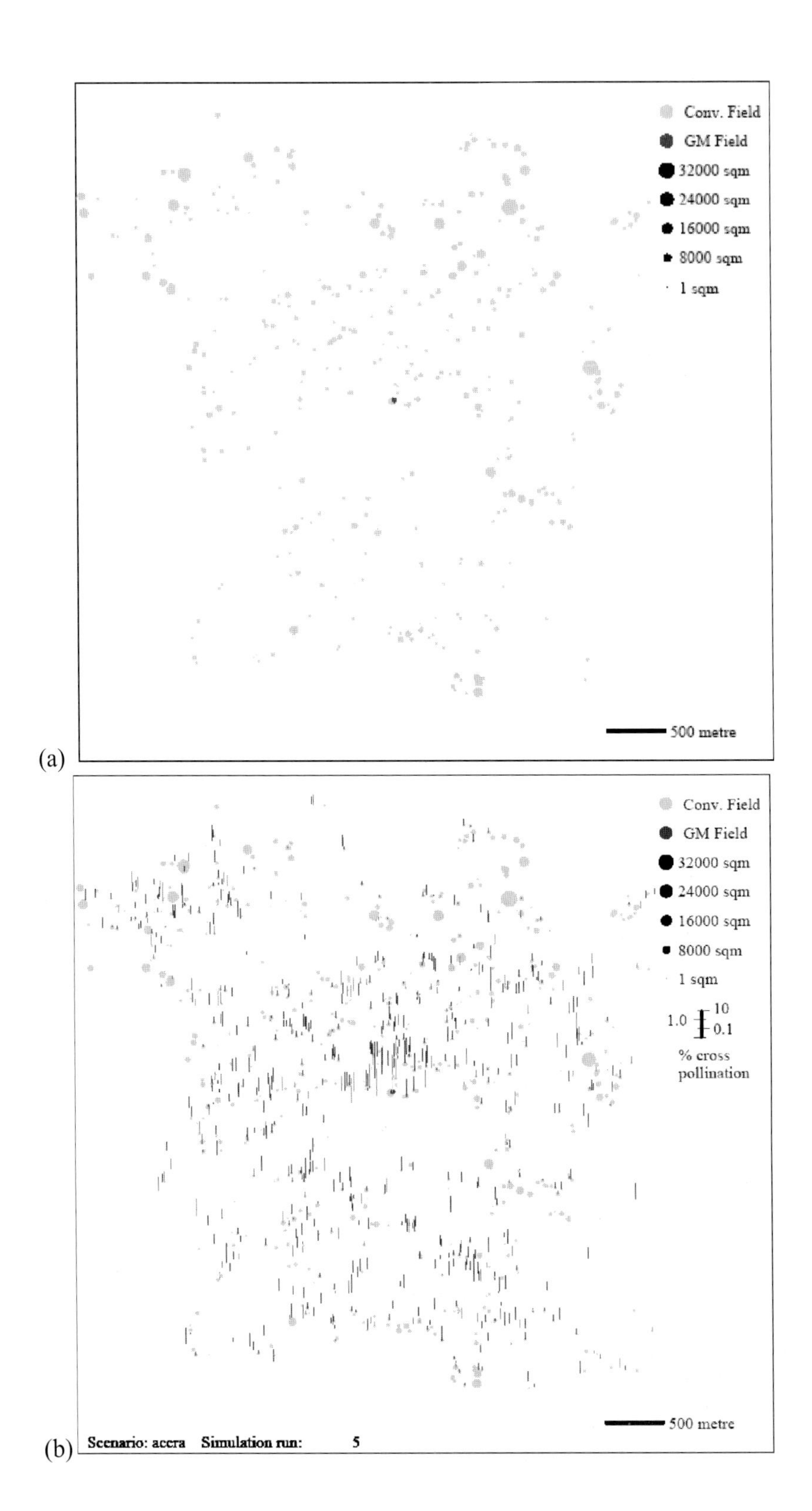
Conv. Field
GM Field
32000 sqm
24000 sqm
16000 sqm
8000 sqm
1 sqm
500 metre
(a)
Conv. Field
GM Field
32000 sqm
24000 sqm
16000 sqm
8000 sqm
1 sqm
1.0
10
0.1
% cross pollination
500 metre
Scenario: accra Simulation run: 5
(b)

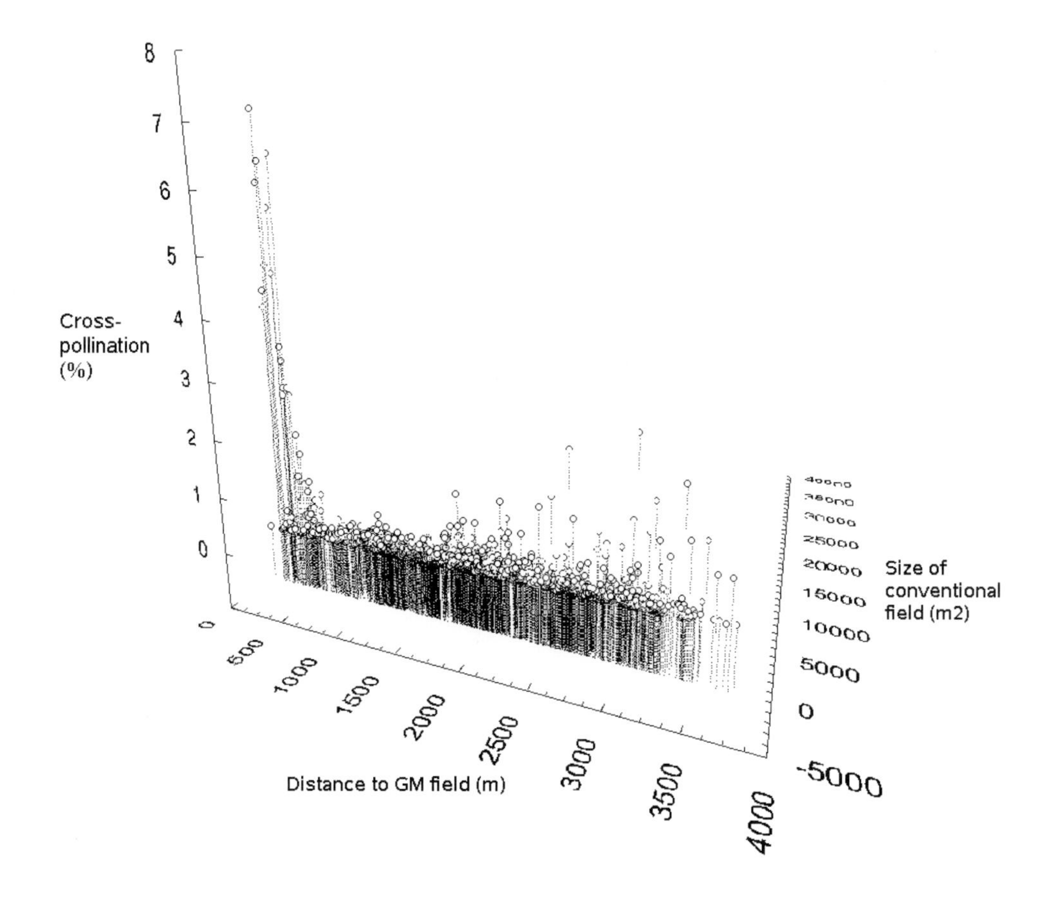

Figure 6.3.16: Percentage cross-pollination rates in 1,389 conventional fields from a single central GM source field as shown in the model Fig. 6.3.15. This 3D display relating cross-pollination, area of the conventional field with its distance to the single GM field reveals a hyperbolic function. For example a conventional field at a distance and size of 300 m (200 m2), 800 m (4,000 m2) and 1500 m (10,000 m2) have average cross-pollination rates of 4.5%, 1.0 % and 0.5% respectively from the single centre GM field. This suggests that pollination rates in conventional fields generally decreases with increase in field size and distance to the GM field.

The results of the simulation model could be compared to those of the GIS approaches as a means to validate the findings (Fig. 6.3.17). Both scenarios show that in the presence of genetically modified crops, larger numbers of neighbouring fields share relatively high cross-pollination rates.

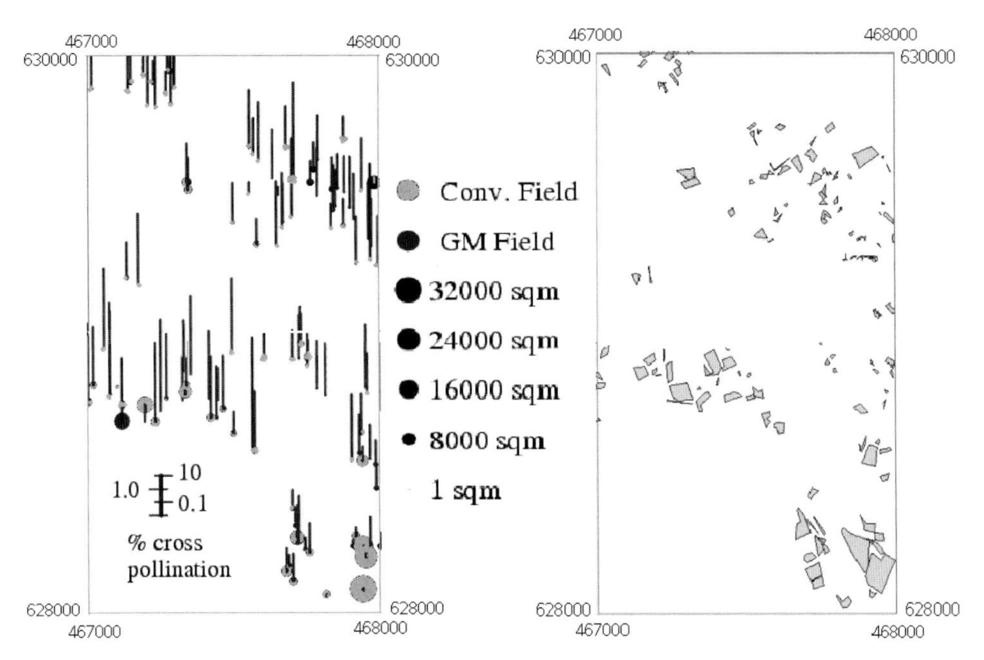

Figure 6.3.17: Validation of model (a) and GIS (b) approaches. Both methods could be used to justify the specific influence of different factors such as size and precise location of conventional fields relative to the GM field on the emerging variability of cross-pollination rates (The arrow on the right map points to the assumed GM field on the left map).

6.3.2 Summary of results from modeling approaches

A summary of scenario calculations is presented below (Table 6.8). For the different scenarios, average presence of GM maize kernels in the harvest are given according to the simulation, the assumed number of GM fields of total fields and % GM area of total maize cultivation area. The table is completed by the results obtained for additional scenarios that were calculated: It was also investigated, how the entry of GM through saved seeds would influence cross-pollination, and how a selection of larger fields for GM cultivation would influence cross-pollination (all fields between 0.5 and 1 Hectare as GM or all fields above 1 ha size GM.

These data can help enforcement authorities to decide on certification standards for GM food or feed products in Ghana with implications for food and feed producing regions and producers. In the EU, GM labeling is mandatory if the threshold of 0.9% is exceeded (EU Regulation 1829/2003 on labeling of genetically modified food and feed). The overall data shows that even though a large number of small fields under certain given conditions have a relatively pronounced heterogeneity in planting times, different flowering periods, and relatively wider field distances apart, it is still possible that relatively small fractions of GM introgression rates in conventional fields in the order of 5% or even below to about 1.0% is sufficient to contaminate the entire cropping region.

Table 6.8: Gene flow conditions based on model approach.

Scenarios	Maize hybrids			Open pollinated varieties		
	Average presence of GM in conven-tion-al harvests	% of GM fields of total fields	% of GM field area of total field area	Average presence of GM in conven-tion-al harvests	% of GM fields of total fields	% of GM field area of total field area
(1) GM planted obtained under exchange conditions	0.87	3.02	2.27	1.41	3.02	2.27
(2) GM planted obtained from food market	1.12	2.52	3.78	1.88	2.52	3.78
(3) GM planted obtained from seed market	2.61	7.70	6.02	4.25	7.70	6.02
(4) GM planted obtained from saved seeds	1.00	4.24	2.40	1.68	4.24	2.40
(5) GM in fields between 0.5-1 Ha	0.27	0.14	0.74	0.42	0.14	0.74
(6) GM in fields greater than 1 Ha	0.46	0.14	2.01	0.64	0.14	2.01
(7) Single GM center field	0.12	0.07	0.21		-	

6.4 Socio-economic Aspects

6.4.1 Demographic profile of respondents

In all, two hundred and one (201) interviews were conducted involving 196 smallholder farmers within a spatially coherent area comprising 13 sub-communities of the Ga West district of Accra.

(a) Age: The age distribution of respondents is shown in Fig. 6.4.1. Nearly 54% of farmers interviewed have mean age range of 25-35 years old representing the active age group of the sample population. 7% of respondents have the lowest mean age of about 15 years. The highest mean age was found to be 75 years making up about 3% of all respondents interviewed. In Fig. 6.4.2, further demographic differences are presented.

(b) Gender: Basic demographic information about survey respondents appears in Figure 6.4.2. The data shows that: About 35% of survey respondents were women. However, several other women farmers were remotely observed working on the farms of their husbands who were informants. This suggests that actual number of women farmers were probably concealed. Male respondents constituted 65% of the sampled population.

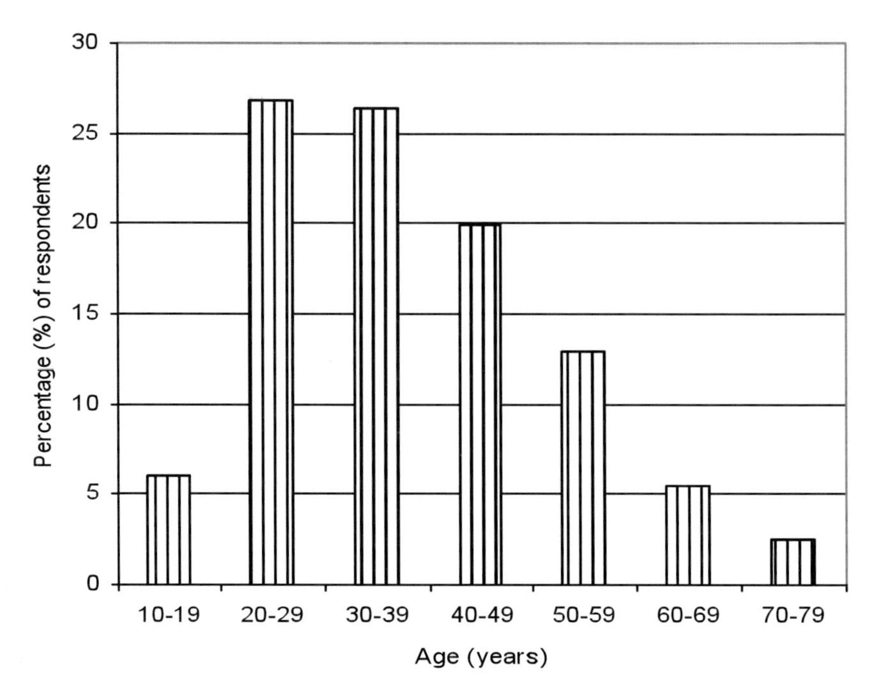

Figure 6.4.1: Age distribution of survey respondents.

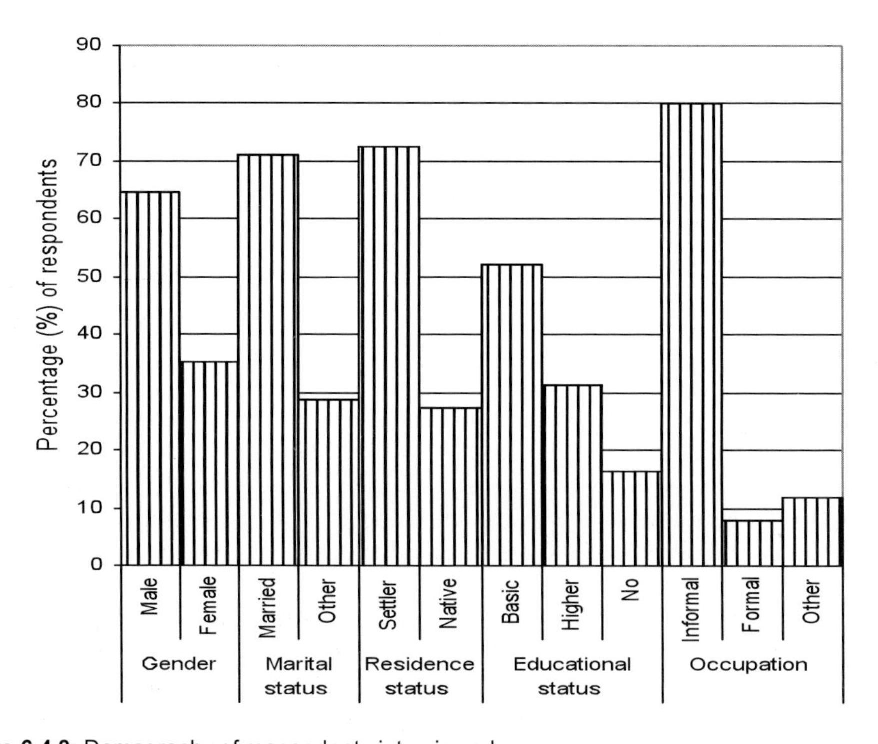

Figure 6.4.2: Demography of respondents interviewed.

(c) Marital status: Three-quarters of the population (71%) were married and the remaining 21% comprising single households.

(d) Residence status: It is also shown that majority of maize farmers interviewed (73%) are settlers with the native population comprising 27% of the population.

(e) Educational status: Analysis of educational status revealed the following:
Slightly more than half of the population (53%) had only basic (primary) education. Those with higher education (mainly secondary and tertiary) constitute 31% of population. While respondents with no education constitute 16.4%.

(f) Occupational status:
An estimated 80% of the population was revealed to be working in the informal sector of the economy e.g. as carpenters, masons, traders or self-employed. Those in the formal sector constitute 8%, made up of teachers, and other recognized professionals. Students and retired people made up 12% of the population.

From Pearson's correlation analysis (Table 6.9), the following causal observations could be made:

- Both men and women are largely in the informal sector occupations with mostly men in the formal sectors;
- The male farmers mostly have marital status unlike women farmers;
- Most of the respondents with marital status are mostly natives suggesting a single status of settler farmers.

6.4.2 Classification of farmer resource capacity at the household level

(a) Household size: The number of persons per househould is shown in Fig. 6.4.3. It shows that about 48% of farmers have an average household size of 1-3 persons, nearly 36% have 4-6 persons on average, 12% have 7-9 persons- while about 6% have more than 9 persons in their households.

(b) Labour capacity: About 68% of farmers have 1-2 persons on avarage to help with on-farm activities. Respondents with highest number of farm hands represent calculates to about 1%, and the suggestion that majority of farms (ca. 93%) are managed largely by the owners and their families.

(c) Land tenure: Nearly 88% of all respondents cultivate maize on privately-owned lands. Individuals who cultivate on leaseholds and public lands constitute 5% and 7% respectively.

(d) Right of access to cultivation: Care-takers constitute 89% of the population, while farmers who own the land cultivated and those with no rights constitute about 5.5% respectively.

Table 6.9: Analysis of population demographic characteristics using Pearson's correlation coefficient. (Marked correlations (*) are significant at $p < .05000$).

Population variables	Male	Female	Married	Other e.g. single. widowed	Native	Settler	Basic education	Higher education	No educa-tion	Informal sector occupation	Formal occupa-tion	Other occupation e.g. retired
Male	1.00											
Female	0.92	1.00										
Married	0.98*	0.89	1.00									
Other e.g. single. widowed	0.72	0.87	0.60	1.00								
Native	0.99*	0.95*	0.98*	0.74	1.00							
Settler	0.98*	0.98*	0.95	0.82	0.99*	1.00						
Basic education	0.90	0.68	0.93	0.37	0.87	0.81	1.00					
Higher education	0.74	0.92	0.65	0.98*	0.78	0.85	0.38	1.00				
No education	0.84	0.98*	0.83	0.84	0.90	0.92	0.58	0.92	1.00			
Informal sector occupation	0.99*	0.96*	0.97*	0.77	1.00*	1.00*	0.86	0.80	0.89	1.00		
Formal sector occupation	0.97*	0.93	0.99*	0.65	0.99*	0.96*	0.89	0.71	0.89	0.98*	1.00	
Other occupation e.g. retired	-0.66	-0.35	-0.74	0.03	-0.61	-0.51	-0.92	0.02	-0.25	-0.59	-0.67	1.00

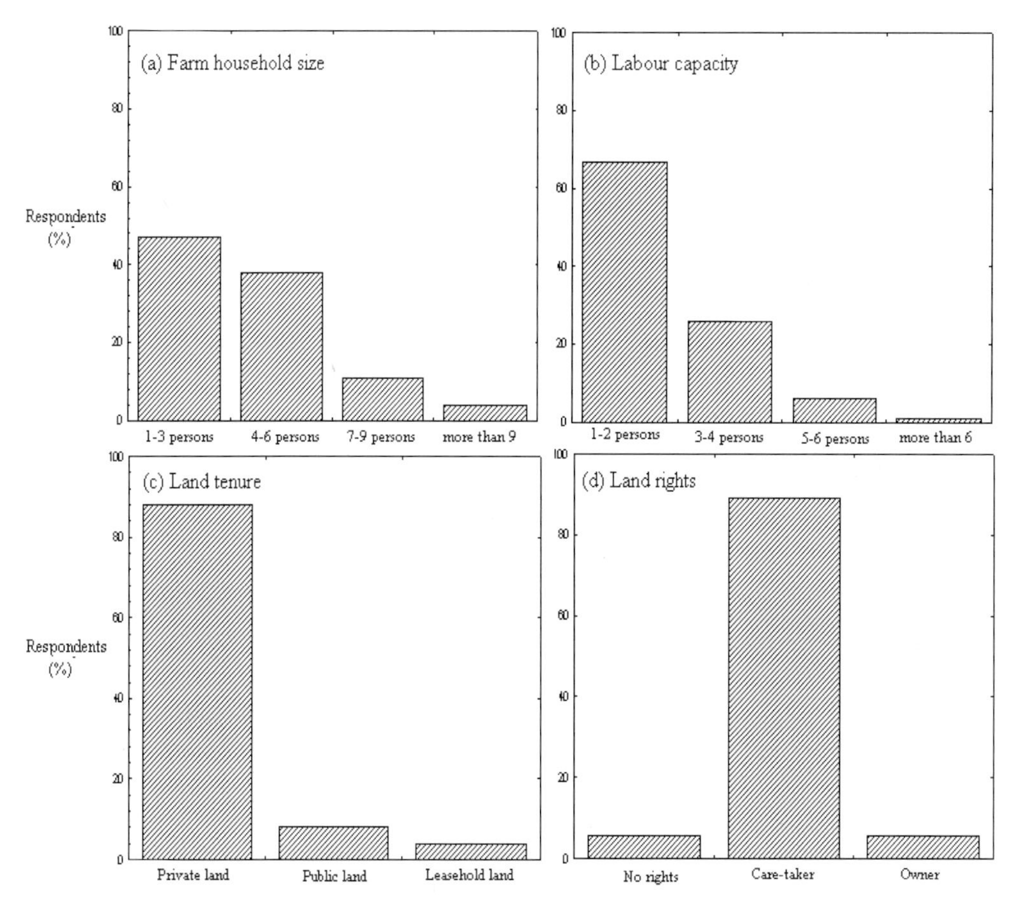

Figure 6.4.3: Farm household structure (a) Household size: Majority of the respondents (45%) comprise about 1-3 persons, and 5% of respondents have a household size of more than 9 persons; (b) Labour capacity: In terms of available labour per household, the data shows that 68% of respondents have 1-2 persons available for farm work, and about 2% of all respondents have about 6 persons or more available; (c) Land tenure: 88%, 8% and 4% of all respondents cultivate their crops on family or private lands, public and leasehold lands respectively; (d) Land rights: An estimated 90% of all respondents cultivate lands as care-takers, while 5% of the population have no rights to land. An additional 5% actually own the land they cultivate.

(e) Technology access: In Fig. 6.4.4., the following are addressed:

- Land preparation: Involves the use of hand tools (nearly 99%) or tractor (1%) and the crop is planted either as a sole crop or intercropped with non-tubers or/ and tuber crops. Regardless of whether it is solely planted or intercropped, maize farms are generally weeded by hand.
- Fertility management: Most farmers apply green manure resulting from on-farm weed clearing as a main source of soil fertility (ca. 53%) or apply nothing at all (40%). Application of inorganic fertilizers appears minimal (7%).
- Pesticide application: 96% of farmers do not apply pesticides while and estimated 4% apply pesticides.

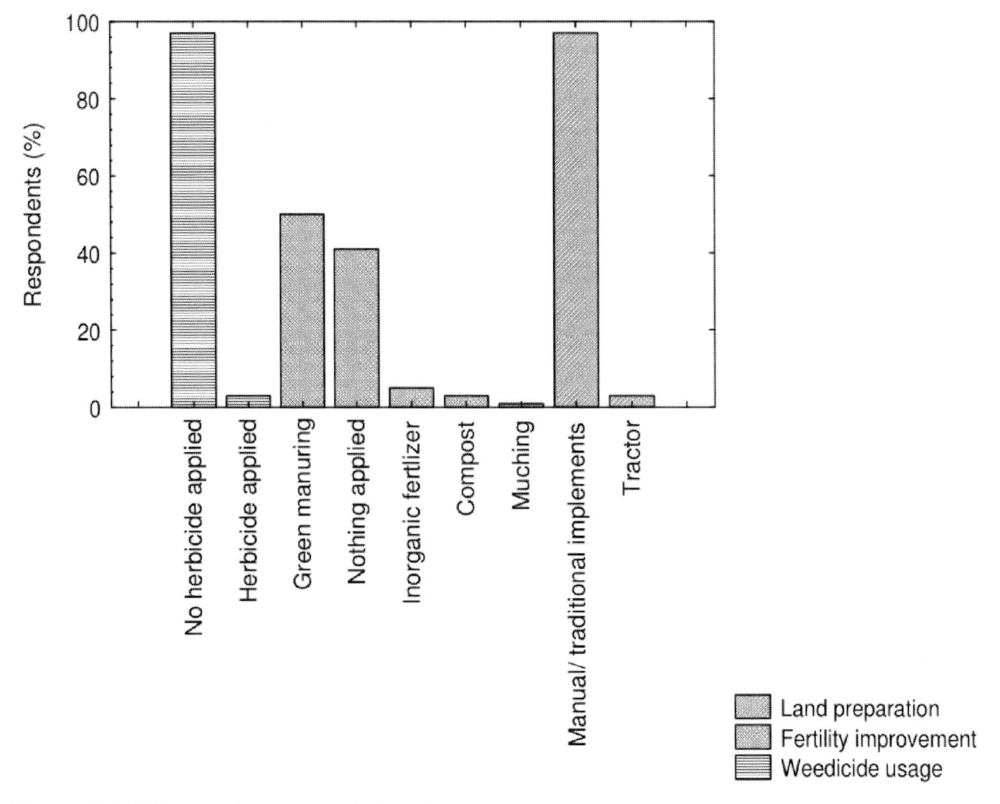

Figure 6.4.4: Farmers' access to technology.

6.4.3 Gender effects and farmers' socioeconomic circumstances

Table 6.11 presents some quantitative indicators calculated for two sub-groups within the survey sample: Male farmers on the one-hand, and female farmers on the other. T-tests are performed to estimate the level of statistical differences if any, between the two populations.

(a) Farmer characteristics: The mean age of male respondents does not differ significantly from that of women respondents. The men earn slightly higher income than the women with a figure of about US$ 2.03. However, the overall household monthly income is higher in the female households. Female farmers have a higher household size as compared to their male counterparts by one person. On the whole the women farmers have been living in the community for an average of 7.6 years as compared to the men with 5.8 years. However the difference is not statistically significant. Men have been growing maize on their lands slightly longer by a year when compared to women counterparts.

(b) Residence status: The male settler population are significantly higher in number than the women. On the other hand, even though the male natives are more than the native female population, the difference is not statistically significant.

(c) Occupation: In terms of occupation, it is deduced that the male population exceed the female population in all the three sectors of the economy, namely the informal, formal and applies also to students and retired people. There are significantly more males in the formal sector as compared to females. The contrary applies in the informal sector. There are significantly higher male students and retired people than females.

(d) Farm orientation: More men are involved in both the subsistence and commercial farming contexts than the women. The differences are statistically significant.

(e) Education: Male farmers have a significantly higher basic and higher education than the women. There are more uneducated women within the sample population as compared to men. The difference in the latter is statistically significant.

(f) Land tenure: Male farmers cultivate significantly more privately owned lands as compared to the females. The data shows that no woman has a leasehold access to land. However is shown that most publicly-owned lands are cultivated by women. This differs significantly from men's cultivation of public areas.

A background analysis of specific intra-household factors using Pearson's correlation analysis (Table 6.10) reveals that:

- Female farmers are mostly in the informal sector of the economy;
- Females are mostly preoccupied in the subsistence sector;
- Household income strongly correlates with availability of labour or presence of a working class within households;
- Subsistence and commercial farming are part of the usual routine;
- Respondents within the formal sector cultivate maize for subsistence, and crop cultivation as a commercial activity is unlikely.

6.4.4 Farmers' seed selection criteria

Table 6.12 specifies the general criteria used by urban farmers for the selection of seeds from various respective sources. An analysis of their level of importance is shown based on the frequency (SD±) of responses by farmers. The specific criteria include the following: High yielding, stable yielding, insect resistance, low input costs, and early maturing varieties. Basing on Table 6.4.4, it is revealed from the farmers' perspective, the following decreasing order of relevance:

- High yielding
- Early maturing
- Low input costs
- Insect-resistance
- Stable yielding

Table 6.10: Analysis of intra-household factors using Pearson's correlation coefficient (Marked correlations (*) are significant at $p < .05000$).

Household variables	Male farmers	Female farmers	House hold size	Average household income	Informal sector occupation	Formal sector occupation	Other occupation e.g. retired. student	Resource capacity. i.e. No. of farms	Labour capacity	Subsistence orientation	Commercial orientation	Period of stay in community
Male farmers	1.00											
Female farmers	0.92	1.00										
Household size	-0.13	-0.48	1.00									
Average household income	-0.03	-0.11	0.51	1.00								
Informal sector occupation	0.99*	0.96*	-0.23	-0.04	1.00							
Formal sector occupation	0.97*	0.93	-0.13	0.17	0.98*	1.00						
Other occupations e.g. retired. student	-0.66	-0.35	-0.65	-0.46	-0.59	-0.67	1.00					
Resource capacity. i.e. No. of farms	0.63	0.32	0.67	0.50	0.56	0.65	-1.00	1.00				
Labour capacity within household	-0.21	-0.27	0.53	0.98*	-0.22	0.00	-0.34	0.38	1.00			
Subsistence orientation	0.99*	0.96*	-0.23	-0.04	1.00*	0.98*	-0.59	0.56	-0.22	1.00		
Commercial orientation	0.97*	0.93	-0.13	0.17	0.98*	0.65	-0.67	0.65	0.00	0.98*	1.00	
Period of stay in community	0.13	-0.19	0.91	0.75	0.05	0.20	-0.82	0.85	0.72	0.05	0.20	1.00

Table 6.11: Analyses of gender and farmers' socioeconomic circumstances.

Factor	Male farmers	Female farmers	Significance level of difference*
(a) Farmer characteristics:			
Age (years)	37.4	37.7	NS
Monthly per capita income (US$)	63.17	61.94	NS
Monthly household income (US$)	243.68	295.57	NS
No. of persons in household	3.9	4.5	NS
Period of stay in community (years)	5.8	7.6	NS
Years of growing maize on land	3.6	2.6	NS
(b) Residence status (% farmers):			
Settler	67	33	<0.001
Native	58	42	NS
(c) Occupation (% farmers):			
Informal	59.6	40.4	<0.001
Formal	68.8	31.2	<0.05
Other (e.g. student. retired)	95.8	4.2	<0.001
(d) Farm orientation:			
Subsistence	90.0	10.0	<0.001
Commercial	97.2	2.8	<0.001
(e) Education (% farmers):			
Basic (e.g. primary)	70.5	29.5	<0.001
Higher (e.g. secondary, tertiary)	71.9	28.1	<0.001
No education	31.3	68.7	<0.01
(f) Land tenure (% farmers):			
Private/ Family	65.3	34.7	<0.001
Leasehold	100.0	0.0	-
Public/ Usufruct	37.5	62.5	<0.05
No. of farms owned	1.8	1.4	<0.01

*T-test (Test for equality of means)

Table 6.12: Farmers' ascribed criteria for seed selection.

S/n Criteria	Mean Number of responses	Standard deviation	Frequency of responses	Percentage (%) responses
(1) High yielding	29.4	38.5	147	28.5
(2) Early maturing	27.6	39.9	138	26.8
(3) Low input costs	18.2	10.4	91	17.7
(4) Insect-resistance	15.2	19.5	76	14.8
(5) Stable yielding	11.6	15.4	58	11.2
(6) No criteria	1.0	1.4	5	1.0

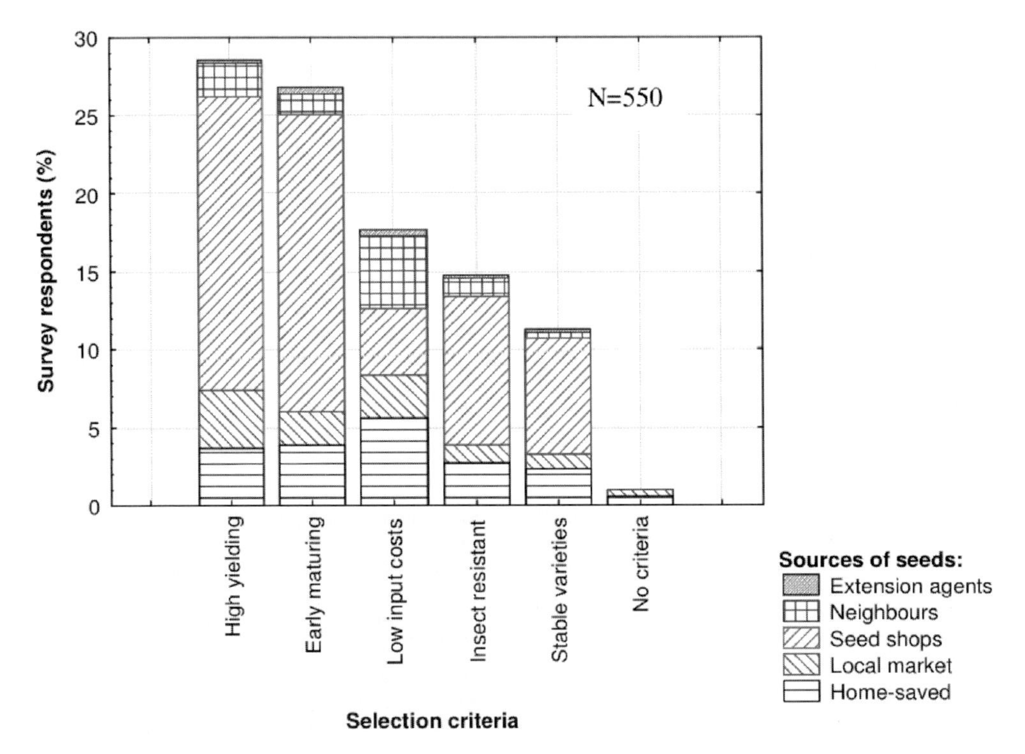

Figure 6.4.5: Perceived criteria for seed selection among sample population. Total number of responses exceeded total number of respondents because of multiple responses. Hence the result is presented in percentage responses, totaling 100%.

From Fig. 6.4.5, the following are deduced:

(a) High yielding: Most farmers go to the shops to acquire commercial seeds for the main reason that they are perceived to be of highest yielding potential. Home-saved and seeds bought from local market are widely used in this context on the assumption of experiences from previous harvests or seed quality. The high yielding factor is least considered in the case of extension agents probably because of the limited activities of extension officers in the region.

(b) Early-maturing: Again, early maturing varieties are procured from shops (inbred lines), and based on early perceived maturity period of previous harvest. Seeds obtained from local market are used in the context probably on a basis of perceived notion of grain features e.g. texture, etc.

(c) Low input costs: Seeds reproduced from previous harvests are mostly used in this context followed by acquisitions from neighbours. Grains obtained from the local market for food, part of which is extracted for sowing appear to be the next relevant criteria. Acquisitions from extension agents constitute the least important criteria.

(d) Insect resistance: Majority of the farmers obtain commercial seeds if they desire specific insect resistance properties of sown seeds. The next important criteria are those from previous harvests.

(e) Stable varieties: This is a nutritional requirement factor. This mirrors the properties of insect-resistance criteria with a higher procurement from formal shops in respect of specific nutritional trait, followed by what is known of the nutritional benefits from previous harvests. In conclusion, the cost factor seem to be the major limitation to procure improved inbred lines from commercial enterprises even though farmers appear to be very much aware of their potential benefits as shown in Figure 6.4.5. The data suggest that locally reproduced and exchanged seeds are crucially relevant and innovative traditional approaches to agriculture and food production in urban areas furthering diversity.

6.4.5 Analysis of seed acquisition and farmers' selection criteria

Judging from the magnitude (and significance levels) of the F values derived from multivariate analyses of seed sources related to selection criteria (K-means clustering) (Table 6.13). It is revealed that local market constitutes the most important source of seed acquisition by local farmers. These are largely obtained primarily for food and some amounts subsequently extracted for planting. Home-saved seeds constitute the second most important source, followed by seeds bought from commercial shops. Seeds obtained from neighbours refer to the least significant seed acquisition source. Seeds procured from extension agents appear hardly possible.

6.4.6 If seeds are bought, are hybrids eventually re-planted?

Analysis reveals that 68% of the farmers replant hybrids due largely to:

- Financial reasons (40% of respondents);
- No ascribed reasons (14% of respondents), suggesting that seed reproduction from previous harvest part of the usual routine for this group;
- Maintenance of specific genetic trait e.g. good tasting etc (6.7%);
- Early maturing, insect resistance and high yielding varieties constitute 4%, 2% and 1.5% of respondent's preferences for reproducing seeds from previous harvests.

Thirty-two (32%) of the farmers who do not re-plant hybrids, and do so for the following key concerns:

- High purchase frequency of grains for food and easy access to seeds (12.4% of respondents);
- Poor yields from previous harvests (8.5% respondents);
- Early harvesting of cobs (green harvesting) for food or for sale (3% respondents);
- Low harvest output, hence harvestable quantities used for food (1.9%)
- Crop rotation measures leading to planting of other crops in subsequent seasons (1.5%) among others (See Table 6.14).

Table 6.13: Multivariate analyses of seed acquisition in relation to farmers' selection criteria in urban areas.

No.	Seed sources	F (Fisher's coefficient)	Significant P*
(a)	Home-saved	47.6045	0.108229
(b)	Local grain market	100.5004	0.074658
(c)	Shops	32.5569	0.130609
(d)	Neighbours	11.2917	0.219169
(e)	Extension agents	-	-

*ANOVA (P=0.05) (K-means cluster analyses)

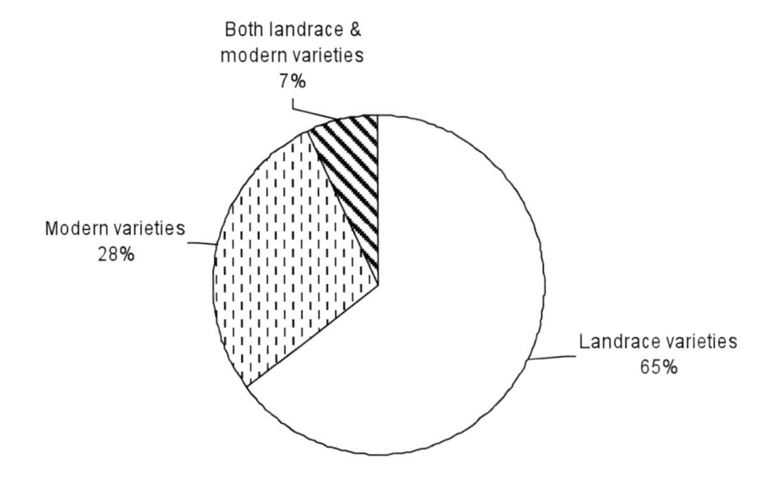

(a) Male respondents

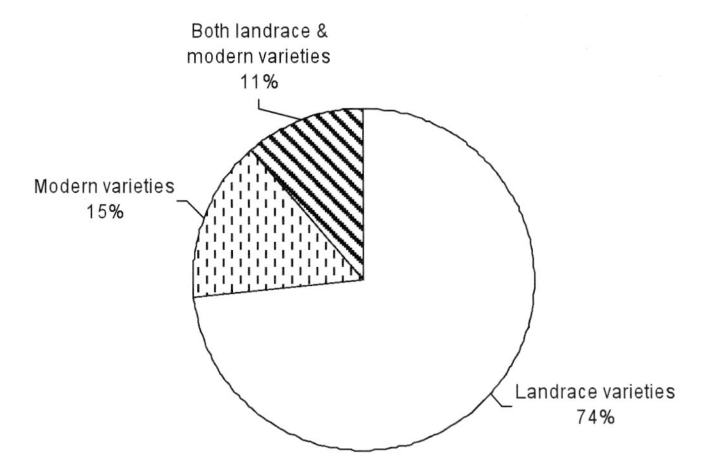

(b) Female respndents

Figure 6.4.6: Modern and landrace seed procurement by (a) male and (b) female respondents in the study area. The data suggests that the wider majority of men and women farmers largely grow maize landrace varieties. In addition, seed mixing involving commercial (modern) and landrace varieties is an accepted practice among a minor section of farmers.

Table 6.14: Reasons for replanting hybrid seeds from previous harvests.

Ranking: Plant hybrids	Respondent's reasons for replanting hybrids	Freqency	% Responses	Cumulative % Responses
(1)	Financial constraints to acquire new seeds	80	40.0	40.0
(2)	No special reasons	28	14.0	54.0
(3)	Favourable genetic trait e.g. nutritional value	13	6.5	60.5
(4)	Early maturing	8	4.0	64.5
(5)	Insect resistance	4	2.0	66.5
(6)	High-yielding	3	1.5	68.0
Ranking: Do not plant hybrids	**Respondent's reasons for not replanting hybrids**	**Freqency**	**% Responses**	**Cumulative % Responses**
(1)	Accessibility to grain markets and formal seed shops	25	12.4	12.4
(2)	Poor yields of previous harvests	17	8.5	20.9
(3)	Avoidance of seed infestation problems	6	3.0	23.9
(4)	Early harvests/ Fresh (green) cobs harvesting	6	3.0	26.9
(5)	Small farm/ Low amount of harvests	4	1.9	28.8
(6)	Planting other crops other than maize in subsequent seasons	3	1.5	30.0
(7)	Seeds always available from quantities bought for food	2	1.0	31.3
(8)	Unfavourable trait	1	0.5	31.8
(9)	No special reasons	1	0.5	32.0
Total		201		

6.4.7 Cultivation period and seasonality:

Planting of maize begins in March involving about 10% of the population (Fig. 6.4.7). However, the number of growers tends to increase in the subsequent months of the rainy season to 27% by April. Peak planting period is recorded in May involving about 31% of the population. Thereafter, significant reductions in crop planting are observed by July involving about 6.5% of farmers. Prior to the minor season in September, some farmers begin to grow maize in August estimating to 3.4% and increases to 9.5% in September when planting ends.

First harvest of the major season plantings begins as early as May involving 0.5% of farmers, increasing to 8.5%, 19.0% and 38.8% in June. July and August respectively.

Lowest harvest period of the major season is recorded in September involving about 18% of farmers. Harvests of the minor rainy season start in October with 2% of farmers continuing into November (4% farmers) and close in December as the highest peak period of the season involving 9.5% of farmers. There is hardly crop production during the extensive dry period from December ending February. Fig. 6.4.6 shows data on seasonal maize food production in the study area.

6.4.8 Demographic factors associated with farm type orientation

Table 6.15 presents data on demographic factors associated with specific farm type orientation. Data are presented in a form of qualitative indicators calculated for two subgroups within the survey population namely: farmers who grow mainly for subsistence purposes and those who grow for commercial reasons. T-tests are performed to determine the level of statistical significance, if any, between the observed differences in the indicators between the two groups.

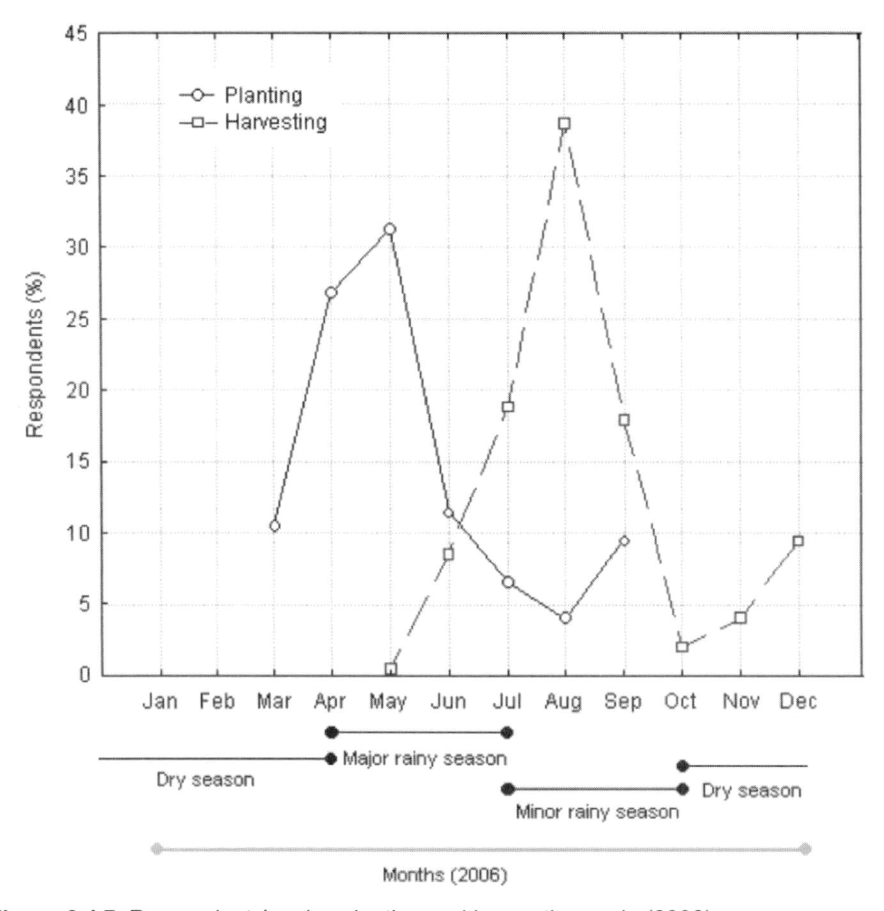

Figure 6.4.7: Respondents' maize planting and harvesting cycle (2006).

(a) Farmer characteristics: Commercially-oriented farmers on the average are older than subsistence-oriented farmers by a margin of about 6 years. However, the difference is not statistically significant. Commercial farmers have a significantly higher household sizes compared to their subsistence counterparts. They also have a higher monthly household income compared to subsistence farmers. However, do not differ significantly. Commercial farmers have been living in the communities for significantly longer periods than subsistence persons. It is also revealed that the commercial farmers have a longer maize cropping history however not statistically significant to that of subsistence farmers.

(b) Residence status: Both natives and settler populations show a significantly higher subsistence practice as compared to their commercial farm enterprises.

Table 6.15: Analyses of demographic factors associated with subsistence and commercial farming approaches.

Factor	Subsistence orientation	Commercial orientation	Significance level of difference*
(a) Farmer characteristics:			
Age (years)	37.0	42.8	NS
No. of persons in household	4.0	5.7	<0.05
Monthly household income (All sources) (US$)	257.9	311.78	NS
Period of stay in community (years)	6.1	10.3	<0.05
Years of growing maize on land	3.2	3.9	<NS
(b) Residence status (% farmers):			
Settler	94.5	5.5	<0.001
Native	91.8	8.2	<0.001
(c) Occupation (% farmers):			
Informal	93.2	6.8	<0.001
Formal	100.0	0.0	-
Other (e.g. student. retired)	83.3	16.7	<0.001
(d) Land tenure/ Resource ownership (% farmers):			
Private/ Family	93.8	6.2	<0.001
Leasehold	77.8	22.2	<0.05
Public/ Usufruct	87.5	12.5	<0.001
No. of farms owned	1.6	2.1	<0.05
(e) Education (% farmers):			
Basic (e.g. primary)	92.4	7.6	<0.001
Higher (e.g. secondary, tertiary)	92.2	7.8	<0.001
No education	93.8	6.2	<0.001
(f) Marital status:			
Married	92.3	7.7	<0.001
Other e.g. single. widow	93.1	6.9	<0.001

*T-test (Test for equality of means)

(c) Occupational status: Most persons in the informal sector of the economy e.g. carpenter, masons, drivers engage in significantly higher levels of subsistence farming. It is deduced that all persons in the formal sector e.g. working as teachers, etc engage mainly in maize farming as a subsistence activity. None of them engage in maize farming as a matter of commercial interests. Students and retired population engage in farming mainly for subsistence with a high level of statistical significance.

(d) Land tenure/ Resource ownership: Privately-owned lands, leaseholds as well as public lands are significantly utilized within the subsistence contexts, than for commercial purposes. However, irrespective of where they are cultivated, number of farm places owned by commercially-oriented people tends to be significantly higher than what is owned to subsistence farmers.

(e) Educational status: Among the educational factors considered. it is deduced that farmers with no education have a significantly higher engagement in subsistence compared to those of basic and higher education. The relations to commercial interests appear to be an irrelevant factor.

(f) Marital status: Single and widowed farmers engage in more subsistence farming than married counterparts. However, marital status does not seem to have a statistically significant influence on commercial approaches.

6.4.9 Intra-household factors influencing seed type utility

Table 6.16 analyses specific household factors associated with the adoption of traditional (TVs) or commercial varieties (CVs). From a socio-economic perspective, these provide baseline indicators for monitoring and are elaborated as follows:

(a) Household characteristics: Households with larger family sizes (4.1) planted more CVs as compared to smaller families (4.0). The difference however, is not statistically significant. It is also observed that respondents who owned more farms planted significantly more CVs (1.8) than TVs (1.5) on an average ($P<0.05$). The result shows that even though CV adopters earned a higher average income (US$272.30 per month) than TV adopters (US$244.00 per month). the difference is not statistically significant. It is shown that the period of stay in the community is not an important factor in the choice of seed type by farmers.

(b) Farm type: Farmers who cultivate for subsistence purposes plant more TVs than CVs. However, both do not differ significantly. It is noted that commercially-oriented farmers plant significantly greater CVs when compared to TVs ($P<0.01$). Thus the hypothesis that commercially-oriented farmers are more likely to invest in improved varieties is supported. Also, CV adopters utilize significantly higher labour than their TV counterparts ($P<0.05$).

(c) Educational status: Respondent farmers with basic education (primary level) do not appear to discriminate between CVs and TV usage. Individuals with higher education (secondary and tertiary) utilize more TVs than CVs. However, this is not a significant statistical difference. It is also shown that the utility of TVs by informants with no education is significantly higher than CVs (P<0.01).

(d) Land tenure: The analyses has shown that informants cultivating private lands employ significantly higher TVs than CVs (P<0.001). Those utilizing leasehold or public lands grow more CVs. However, the difference is not statistically significant.

(e) Cropping intensity: Cropping intensity is seen to be positively associated with the use of TVs. Maize fields that are cultivated twice (2x) per year and intercropped received more TVs than CVs. not statistically significant. Fields cultivated once (1x) per year and left to fallow as well as those intercropped have shown a significantly higher cultivation of TVs than CVs (P<0.05) respectively.

Table 6.16: Factors associated with farmers' adoption of local and modern varieties.

Factor	Plant traditional varieties (TVs)	Plant commercial varieties (CVs)	Significance level of difference*
(a) Household characteristics:			
Household size	4.0	4.1	NS
Monthly household income (US$)	244.00	272.30	<0.05
Period of stay in community (years)	5.5	7.1	NS
Labour capacity (No. of persons engaged on-farm)	1.9	2.4	<0.05
(b) Farm type (% farmers):			
Subsistence	56.5	43.5	<0.05
Commercial	23.1	76.9	<0.01
(c) Educational status (% farmers):			
Basic (e.g. primary)	50.0	50.0	NS
Higher (e.g. secondary. tertiary)	52.5	47.6	NS
No education	66.7	33.3	<0.001
(d) Land tenure (% farmers):			
Private/ Family	54.5	45.5	<0.001
Leasehold	44.4	55.6	NS
Public/ Usufruct	68.8	31.2	<0.05
No. of farms owned (cultivated)	1.5	1.8	NS
(e) Cropping intensity (% farmers)			
2x/year maize (sole) cropped	54.5	45.5	<0.001
2x/year maize intercropped	50.0	50.0	NS
1x/year maize cropped. then fallow	58.3	41.7	<0.05
1x/year intercropped with tubers	52.4	47.6	NS
1x/year intercropped with non-tuber crops	73.7	26.3	<0.05

*T-test (Test for equality of means)

6.4.10 Farmers land rights

A cluster analysis of Euclidean distances between land ownerships and specific right of access to cultivation is shown in Fig. 6.4.8. 'Leasehold lands' are closely associated with 'ownership rights' (linkage distance of 1.22), suggesting that most farms purported to be 'owned' had been obtained on lease basis for a certain number of years. Those who cultivate maize on 'Public lands' form a cluster with those who have 'no rights' to land (linkage distance of 1.93). This suggests the use of publicly available areas by landless inhabitants for cultivation. There is a significant difference observed between clusters comprising 'Private land and Care-takers' on the one hand and 'Leasehold and Owned lands. The analysis also reveals that private lands are mostly cultivated by care-takers who hold the land in trust for their employers.

6.4.11 Implication of land tenure and land rights on cropping intensity

Land tenancy arrangement seems to be a major factor influencing the level of cultivation intensity ($P<0.05$). This means that tenancy arrangement determines how farmers intensify cultivation. Agricultural intensification seems to favour growers on privately-owned lands accounting for nearly 88% of all respondents (Fig. 6.4.9). Leasehold and Public lands account for about 4% and 8% respectively.

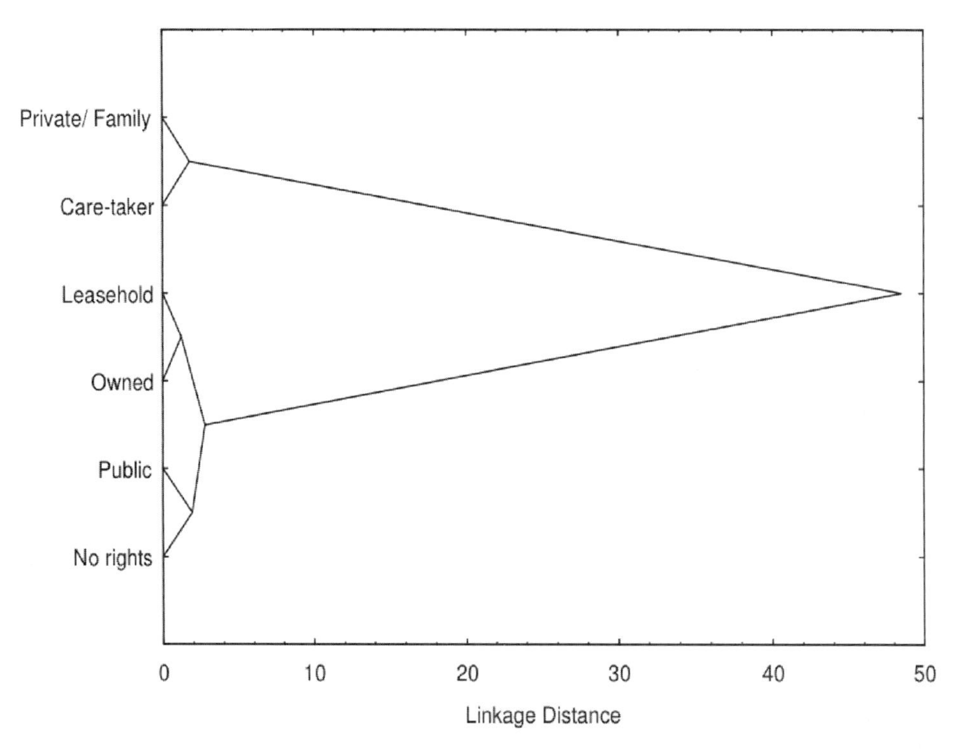

Figure 6.4.8: Tree diagram to estimate the relation between land tenure and specific land rights of farmers.

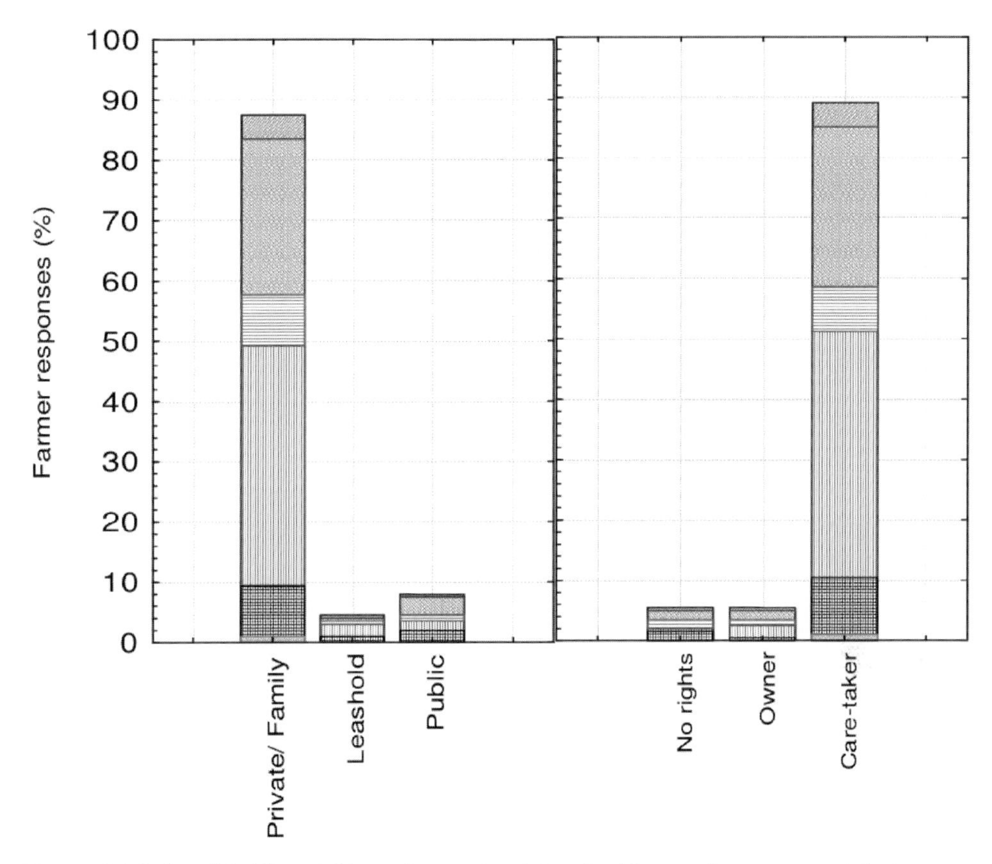

Figure 6.4.9: Land holding rights and cropping intensity. The analyses suggest that resource ownership is a key determinant influencing the level of cropping. However, maize cropping-whether it is done seasonally, continuous or once and fields left to fallow afterwards or grown in association with other crops do not statistically differ from each other. Rights to cultivation is closely linked to levels of cropping intensity ($P<0.05$). This strongly suggests that care-takers have rights to cultivation on privately-owned lands, and comprise 89% of cultivators. Therefore, those with no rights to land cultivate less intensively, just as the case for leaseholders. It is again deduced that the different cropping approaches do not differ significantly among each other. This suggests that the differences in intensity of maize cultivation, whether grown singly or done in association with other crops are not substantial in the small-scale context.

6.4.12 Crop rotational measures

In Table 6.17, an analysis of crop rotational measures is presented. In general, maize growing twice a year as a sole crop is most widely practiced. This is followed by maize growing twice a year however intercropped with other food crops e.g. vegetables. Maize is cultivated once per year (mostly in the major season) if the same cropping area is intended for later cropping of cassava.

Table 6.17: Estimations of crop rotational measures.

Cropping measures	2x/yr maize cropped	2x/yr maize intercropped	1x/yr maize cropped, then fallow	1x/yr inter-cropped with tubers	1x/yr inter-cropped with non-tubers
2x/yr maize cropped	(10) 64.89				
2x/yr maize intercropped	(4) 299.33	(1) 1406.00			
1x/yr maize cropped, then fallow	(11) 47.56	(5) 220.33	(12) 34.89		
1x/yr maize intercropped with tubers	(6) 188.89	(2) 883.33	(7) 138.89	(3) 555.56	
1x/yr maize intercropped with non-tubers	(13) 26.44	(8) 123.67	(14) 19.44	(9) 77.78	(15) 10.89

*COVAR (Covariance) [(1)-(15) Most to least widely used or combined measures].

6.4.13 Analysis of socioeconomic background of farmers in relation to farm land holdings

(a) Maize area per household member: The average maize area available per person per household is shown in Table 6.18. These data are relevant to show how many persons per household depend on a hectare of land as a source of livelihood. There is the indication that larger household sizes do not necessarily lead to larger farm acreages and the suggestion of farmland as a major constraint among larger households. The results show that a household comprising one person has an average field size of about 526m^2. Depending on their socioeconomic situation, this could go up to a maximum of 17,379 m^2 per person per household. Therefore, land holding at the household level is widely variable. For instance, households of about 6 persons have a relatively large average share in terms of farm acreages amounting to 1,379 m^2 while household sizes of about 3 persons have a an average size of 1,298.5 m^2 in terms of area available for cultivation.

(b) Maize acreage available to men and women: Gender-based perspectives on landholding by men and women are important to assess to what extent land could contribute to the socioeconomic development of the local population. Table 6.19 shows that men generally possess higher maize acreages on average when compared to that of women. All male respondents interviewed have an average farm acreage of 920 m^2, amounting to about 1.3 times that which goes to women at 659 m^2. The difference is however not statistically significant. Even though the men have generally larger average farm acreages than the women, the latter have a larger minimum acreage 5 times as much suggesting that the men occupy least marginal areas for cultivation.

(c) Role of marital status on maize land holding: Married men hold more land than unmarried men but on the contrary for the women (Table 6.20). However, in terms of maximum acreage married women tend to have more land than their unmarried counterparts while this situation does not differ for the men.

Table 6.18: Maize acreage at the household level.

Household size (Average no. of persons per household)	Number of cases	Mean acreage (m^2)	Median acreage (m^2)	Total acreage (m^2)	Minimum acreage (m^2)	Maximum acreage (m^2)
1 person	33	526.6	208.0	17379.0	2.0	2523.0
2 persons	24	540.0	328.5	12959.0	4.0	2068.0
3 persons	37	1298.5	358.0	48044.0	3.0	22554.0
4 persons	34	783.3	274.0	26631.0	32.0	4405.0
5 persons	24	526.8	300.0	12643.0	13.0	1961.0
6 persons	19	1379.8	265.0	26217.0	55.0	17668.0
7 persons	9	1196.0	119.0	10763.0	2.0	5892.0
8 persons	10	641.4	350.5	6414.0	42.0	2301.0
9 persons	3	301.3	212.0	904.0	43.0	649.0
> 9 persons	8	561.8	526.5	4494.0	21.0	1271.0
Total	201	829.1	2942.0	166448.0	217.0	61292.0

Table 6.19: Farm size holdings by men and women.

Variable	Men	Women
Number of cases	130	71
Mean acreage (m^2)	920.5	659.0
Median acreage (m^2)	309.5	218.0
Total acreage (m^2)	119660.0	46788
Minimum acreage (m^2)	2.0	10
Maximum acreage (m^2)	22554.0	7000
Standard Deviation	2529.6	1210.2

Table 6.20: Farm size holdings estimated according to marital status.

Variable	Married men	unmarried men	married women	unmarried women
Number of cases	90	40	53	18
Mean acreage (m2)	437.5	212.5	212	241.5
Median acreage (m2)	99661	19999	39860	6928
Sum acreage (m2)	2	2	10	25
Maximum acreage (m2)	22554	2246	7000	1366

Table 6.21: Analysis of period of stay in community on land holding.

Period of stay (years)	Valid N	Mean acreage (m^2)	Median acreage (m^2)	Sum acreage (m^2)	Minimum acreage (m^2)	Maximum acreage (m^2)
<10	160.0	777.7	270.5	124424.0	2.0	22554.0
10-19	27.0	578.8	212.0	15628.0	18.0	2542.0
20-29	5.0	2688.0	1756.0	13440.0	55.0	7000.0
>30	9.0	1439.6	1162.0	12956.0	46.0	3173.0

(d) Implication of settlement history on land holding: An analysis of settlement history on land property holding (Table 6.21) reveals that those who have lived in the community for less than a decade have a generally lower farm sizes but also have the maximum acreage available for cultivation. Respondents who have lived in the community for 3 decades have the highest amount of land available for cultivation. Persons who have lived within the community for more than 3 decades have less acreages least available acreage for cultivation.

6.4.14 Summary results of socio-economic data

The implications for gene flow based on the socio-economic survey are presented in Table 6.22. The analysis point to a highly vulnerable context within the small-scale settings, and therefore, the monitoring for unanticipated adverse effects of GMOs on the social economy would require serious and realistic provisions to enforce. Seed sources are varied, and therefore, traceability to track supply chains for food safety and quality control appear very difficult to implement because of the inherent practices of seed exchange, seed re-use, procurement, land use, crop rotation preferences, and a difficulty of actual product differentiation among small farmers. This is expected to render recall of a potential GM release hardly possible. This further suggests that the challenges are numerous, constrained by the lack of attention on the sector. The data points to the fact that farmers actually constitute a very important source of information due to their direct involvement in cultivation, and the demands to find adequate means of living makes farming on marginal areas very valuable for subsistence in urban areas. Women relate to a key part of the process, implicitly contributing to the overall income and household food security through maize farming. Women are less empowered due to their low education, and the extent of their involvement potentially makes them most liable in an event of adversities.

The next section focuses on the findings of the genomic analysis of the seed samples obtained from local farmers as well as from commercial sources. Attempts were made to develop DNA fingerprints of the local samples of farmer varieties in comparison with commercial varieties as a first step towards interpreting genetic variability or their relatedness, and to open up the relevance of this field in the broader biosafety context in Ghana. The analysis was achieved in the laboratories of the University of Bremen.

Table 6.22: Gene flow conditions based on smallholder socioeconomic and seed exchange factors.

Factor	Descriptor	Gene flow conditions
(1) Seed sources, seed exchange background	For the particular season, farmers growing:	
	home-saved seeds/ reproduced from previous harvests	Highest probable case
	seeds from local grain market	Probable but not likely
	seeds from formal shops/ improved varieties	Very probable via subsequent growing of hybrids
	seeds from home neighbours	Likely but very low
	seeds from extension officers	Likely but extremely low
(2) Gender effects	(a) Educational status of farmers (most significant factors, %):	
	Women (No education)	68.7%
	Men (No education)	31.3%
	Women (Basic education)	29.5%
	Men (Basic education)	70.5%
	Women (Higher education)	28.1%
	Men (Higher education)	71.9%
(3) Seed acquisition, criteria and preferences	(a) Reasons for reproducing seeds from previous harvests by farmers:	
	Income constraint	High concern
	Early maturity	High concern
	Pest sensitivities	Important but low concern
	High yield, nutritional concerns	High concern
	(b) Reasons for acquiring improved seeds from formal shops by farmers:	
	Early maturity	High concern
	Pest sensitivities	High concern
	High yield. nutritional concerns	High concern
	(c) Reasons for seed exchange with neighbours (% farmers):	
	Income constraint	High concern
	Early maturity	High concern
	Pest sensitivities	Important but low concern
	High yield	High concern
(4) Farm-type orientation	(a) Respondents in informal sector (Most significant factors, %):	
	Mainly for subsistence	93.2%
	Mainly for markets	6.8%
	(b) Respondents in formal sector (%):	
	Mainly for subsistence	100%
	Mainly for markets	0%

Factor	Descriptor	Gene flow conditions
(5) Seed variety adoption by respondents	(a) Households growing traditional varieties have: Low household size (ca. 4 persons) Low income levels (ca. $244.00 per month) Lower land rights (e.g. have lower no. of farms on average) High cropping frequency per year (b) Households growing improved varieties have: Low household size (ca. 4 persons) Higher income levels (ca. $272.00 per month) Higher land rights (e.g. have more farms on average) High cropping frequency per year	 Yes, very likely Yes, very likely Yes, very likely Yes, very likely Yes, very likely Likely via subsequent growing of hybrids Yes, very likely Yes, very likely
(6) Cropping intensity and rotation measures	Maize growing (Most significant factors, % farmers): 2x/ year traditional seeds sole cropped 2x/ year commercial seeds sole cropped 1x/ year traditional seeds cropped, then seasonal fallow allowed 1x/ year commercial seeds cropped, then seasonal fallow allowed 1x/ year traditional seeds intercropped with non-tubers 1x/ year commercial seeds intercropped with non-tubers	 54.5% 45.5% 58.3% 41.7% 73.7% 26.3%

6.5 Molecular Genomic Aspects

This section presents initial results of the genomic studies which was executed as a methodological outlook for the Ghanaian biosafety context.

6.5.1 DNA concentration and absorbance measurements

Since samples were collected from different farmers in several locations, they have been considered as different populations as shown by their sampling number (Table 6.23). In all, total genomic DNA yield obtained from all 66 extracts ranged between 272.07-830.23 ng/µl (Popn. 1-60) for open pollinated varieties from local farmers. Those of commercial hybrids (Popn. 61-64) were recorded to be in the range of 276.44- 828.91 ng/µl. The Spanish variety recorded the highest yield of DNA amounting to 1950.08ng/µl. while the exotic variety measured 724.4 ng/µl. However, optical density of all extracts falls within a linear purity range of 0.96- 1.28. See Table 6.5.1 for photometric measurements. Note that samples DWA_13/1- DWA_17/12 are local open pollinated varieties (OPVs) obtained from local farmers. DWA_18/1-18/4 are commercially certified varieties; DWA_18/5 Local Spanish variety and DWA_18/6 is the exotic variety (Date of extraction/absorbance measurements = 11.01- 18.01.2008).

Table 6.23: Photometric measurements of DNA extracts obtained from the maize samples.

No. of popn	Sample ID	DNA Concentration (ng/µl)	Absorbance (A260)	Absorbance (A280)	Purity (260/280)
1	DWA_13/1	257.49	5.150	4.834	1.07
2	DWA_13/2	265.15	5.303	4.590	1.16
3	DWA_13/3	251.17	5.023	4.545	1.11
4	DWA_13/4	341.55	6.831	5.979	1.14
5	DWA_13/5	830.23	16.605	13.878	1.20
6	DWA_13/6	353.41	7.068	6.357	1.11
7	DWA_13/7	276.25	5.525	5.205	1.06
8	DWA_13/8	272.16	5.443	4.933	1.10
9	DWA_13/9	361.25	7.225	6.124	1.18
10	DWA_13/10	304.35	6.087	5.731	1.06
11	DWA_13/11	258.03	5.161	4.889	1.06
12	DWA_13/12	309.92	6.198	5.653	1.10
13	DWA_14/1	260.46	5.209	5.445	0.96
14	DWA_14/2	362.67	7.253	6.621	1.10
15	DWA_14/3	337.23	6.745	5.906	1.14
16	DWA_14/4	415.28	8.306	6.762	1.23
17	DWA_14/5	316.66	6.333	5.633	1.12
18	DWA_14/6	267.99	5.360	5.152	1.04
19	DWA_14/7	508.49	10.170	9.090	1.12
20	DWA_14/8	542.43	10.849	8.928	1.22
21	DWA_14/9	406.86	8.137	6.740	1.21
22	DWA_14/10	426.40	8.528	7.209	1.18
23	DWA_14/11	337.53	6.751	6.129	1.10
24	DWA_14/12	416.33	8.327	7.391	1.13
25	DWA_15/1	304.07	6.081	5.979	1.02
26	DWA_15/2	278.65	5.573	5.024	1.11
27	DWA_15/3	297.56	5.951	5.304	1.12
28	DWA_15/4	536.56	10.731	9.534	1.13
29	DWA_15/5	357.10	7.142	6.931	1.03
30	DWA_15/6	270.24	5.405	5.476	0.99
31	DWA_15/7	299.58	5.992	5.506	1.09
32	DWA_15/8	398.15	7.963	7.027	1.13
33	DWA_15/9	408.67	8.173	7.266	1.12
34	DWA_15/10	248.51	4.970	4.556	1.09
35	DWA_15/11	238.48	4.770	4.474	1.07

No. of popn	Sample ID	DNA Concen-tration (ng/µl)	Absorbance (A260)	Absorbance (A280)	Purity (260/280)
36	DWA_15/12	306.39	6.128	5.355	1.14
37	DWA_16/1	251.07	5.021	4.581	1.10
38	DWA_16/2	336.20	6.724	6.127	1.10
39	DWA_16/3	319.79	6.396	5.470	1.17
40	DWA_16/4	283.69	5.674	4.993	1.14
41	DWA_16/5	298.72	5.974	5.425	1.10
42	DWA_16/6	499.81	9.996	8.362	1.20
43	DWA_16/7	355.90	7.118	6.406	1.11
44	DWA_16/8	380.12	7.602	7.043	1.08
45	DWA_16/9	219.52	4.390	3.809	1.15
46	DWA_16/10	308.03	6.161	5.835	1.06
47	DWA_16/11	352.16	7.043	6.159	1.14
48	DWA_16/12	383.98	7.680	6.804	1.13
49	DWA_17/1	370.31	7.406	6.094	1.22
50	DWA_17/2	236.18	4.724	4.165	1.13
51	DWA_17/3	536.91	10.738	9.580	1.12
52	DWA_17/4	288.64	5.773	5.261	1.10
53	DWA_17/5	343.06	6.861	6.086	1.13
54	DWA_17/6	243.04	4.861	4.571	1.06
55	DWA_17/7	246.50	4.930	4.444	1.11
56	DWA_17/8	301.59	6.032	5.527	1.09
57	DWA_17/9	335.25	6.705	5.987	1.12
58	DWA_17/10	504.53	10.091	7.894	1.28
59	DWA_17/11	298.95	5.979	5.813	1.03
60	DWA_17/12	254.31	5.086	4.706	1.08
61	DWA_18/1	276.44	5.529	5.505	1.00
62	DWA_18/2	283.11	5.662	5.245	1.08
63	DWA_18/3	304.43	6.089	5.604	1.09
64	DWA_18/4	828.91	16.578	13.580	1.22
65	DWA_18/5	1950.08	39.002	32.099	1.22
66	DWA_18/6	724.42	14.488	12.894	1.12

6.5.2 SSR analysis

Fig. 6.5.1 and 6.5.2 presents results of amplification polymorphism of primers phi 331888 and phi 108411 respectively for all local cultivars and commercial varieties. The Spanish and exotic varieties are used for species control (pedigree).

Fig. 6.5.1 (a-c) show differences between the local open-pollinated varieties and commercial varieties, though a full differentiation would require further assessment for which the budget was not available.

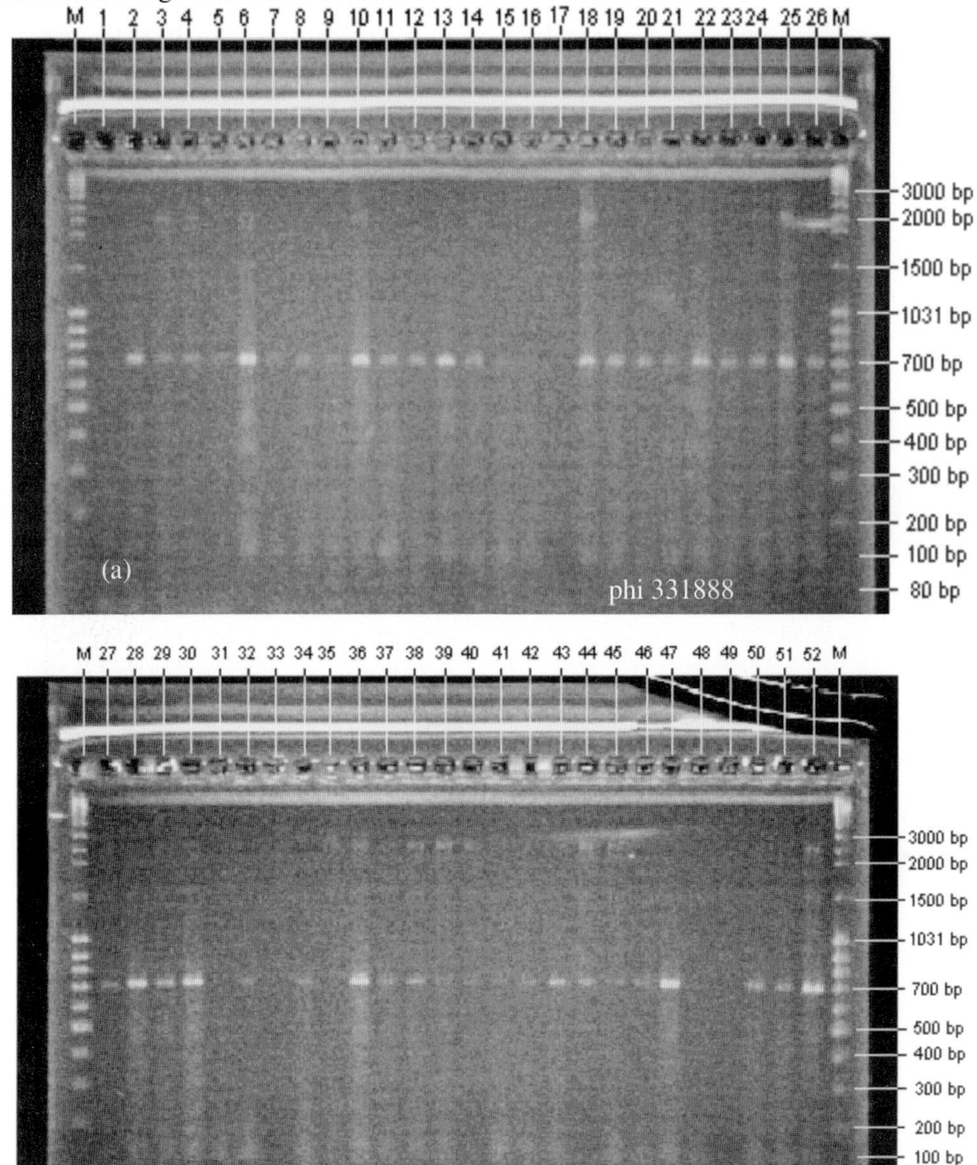

Figure 6.5.1 a & b: Simple Sequence Repeat (SSR) profiles obtained on agarose gel for 49 maize landraces obtained from farmers in Accra. Lane 1 = M-Mass Ruler®; Size; Lane 2-52= Landraces; Lanes 1 & 27 are negative controls.

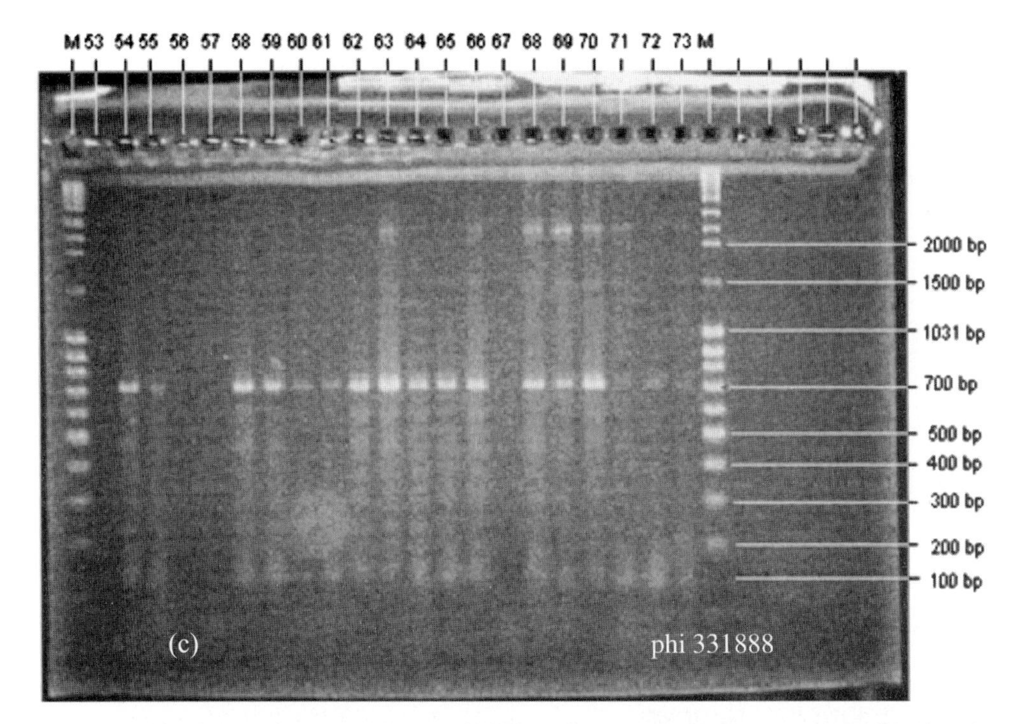

Figure 6.5.1 c: Simple Sequence Repeat (SSR) profiles obtained on agarose gel for 11 maize landraces obtained from farmers in Accra. Lane 1 = M-Mass Ruler®; Size; Lane 53-60= Land-races; Lane 61-64 (Commercial varieties); Lane 65-67 Spanish variety (Seville); Exotic variety (Bremen)

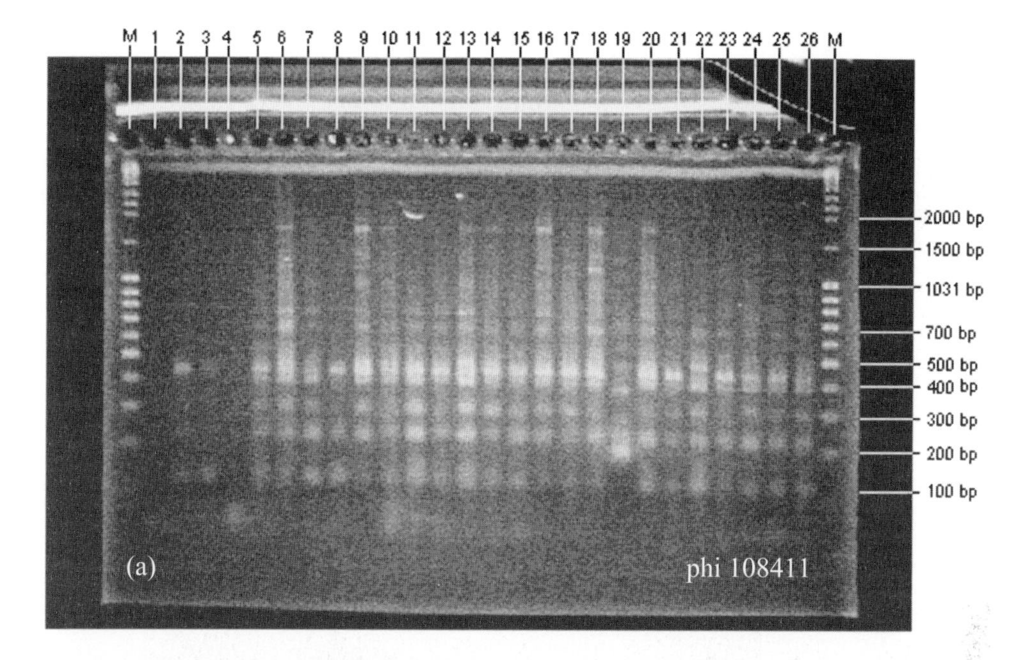

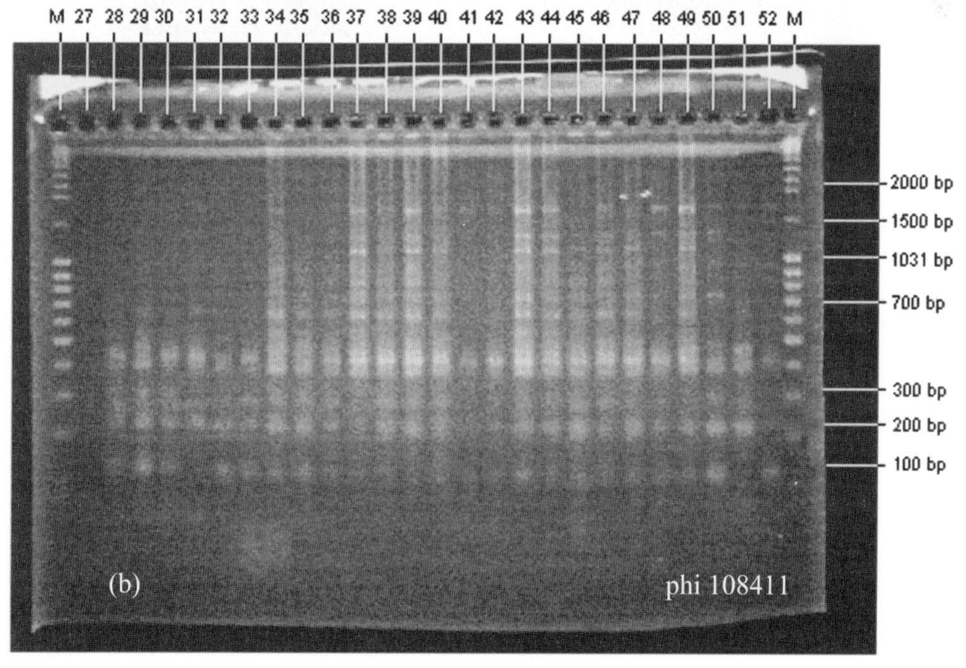

Figure 6.5.2 a & b: Simple Sequence Repeat (SSRs) profiles obtained on agarose gel for 66 populations of Z. Mays (OPVs from local farmers (LVs) and commercial varieties (CVs) from Ghana; local spanish and exotic variety). Lane 1 = Ladder (M-Mass Ruler®); Lanes 2-46= Landraces from local farmers; Lanes 1 & 27 are negative controls; Lanes 47-50 (Commercial varities); Lane 51-52 Spanish variety (Seville) and exotic variety (Bremen).

Chapter 7: Discussion

The overall objective of the study was to provide basic scientific data relevant for biosafety assessment for African agriculture concerning the potential impact of the introduction of genetically modified (GM) maize crops for small-scale subsistence farming. The emphasis here was to analyze for small farm scales, some agroecological and socioeconomic factors that determine whether coexistence of GM and their conventional or organic forms would be feasible for the African conditions. Among others, this was done by assessing the spatial structure of farming, the effectiveness of separation distances of fields as a management measure and to analyze certain underlying agrocultural factors that may influence the extent of gene flow due to pollen movement, seed handling and farmer seed exchange practices. For the African conditions, biosafety data are very scarce. Therefore, it was of crucial importance to analyze baseline data and information for the first time covering various disciplines involving geographical, ecological, and socio-economical structure of cultivation practices of the main cropping system of interest in Accra. In the following, a discussion is made in the light of the results and literature findings relevant for stakeholders and policy-makers to evaluate the options.

7.1 Relevance of the study for the African biosafety context

This study draws attention to relevant conditions necessary for addressing biosafety in Africa and to come up with some possible manageability suggestions with reference to the focal crop of maize. Maize plays a very critical nutritional role across sub-Saharan Africa and is hardly used as a forage crop because it represents an important food staple among the wider part of the population. Due to this, maize landraces represent an important genetic resource that deserves study and active conservation. In the wider global context, GMOs are an issue. The technological options include among others, the creation of plants with specific desired properties for agricultural applications. Genetically modified insect resistant and herbicide tolerant plants are presently the most relevant agricultural biotechnology products on a global scale. There are extensive economic interests behind it mostly driven by the Biotech companies. The introduction of GMOs on the contrary, had brought about intense consumer rejections due to the fact that GMOs have caused practically widespread legal and economic problems elsewhere (Hewlett & Azeez, 2007; Greenpeace, 2007; FoEE, 2007). They have also been advertised to feed the poor countries in Africa as well (e.g. ABSF, 2004).

The context of peri-urban agricultural enterprise in Accra has been used to represent a typical case of small-scale farming in Africa where many farmers engage in various traditional and innovative forms of agriculture with beneficial implications for genetic diversity and sustainable local economies. Urban agriculture is very popular among the migrant population who come from the rural areas in search of making a living in the cities. For this group, it has not only the advantage to generate some small income but also contributes to family and household subsistence. Though being a rapidly expanding and dynamic sector of present African societies, it is largely marginalised and unrecognised by agricultural, agro-ecological and extension research. Its proximity to metropolitan city areas and entry ports brings the expectation that accessibility to imported foreign technologies including food commodities and commercial seed grains might either directly or indirectly introduce genetically modified crop plants into the local environment and agroecosystems. The sector is largely informal and living standards in the urban peripheries are uncertain presenting a difficult livelihood situation for many urban and migrant dwellers.

To overcome these problems, people use open spaces for maize cultivation often in association with other crops regardless whether on small plots in backyards or just cultivate on marginal lands along roads in hope of a harvest someday. Others cultivate on relatively larger portions of property within neighbourhoods even if no formal land rights exist between them and actual landowners. Thus, an understanding of the potential risks involved due to their introduction is very relevant. For regulatory purposes, this study has addressed a broad context as a basis to minimise uncertainties due to potential releases of genetically modified maize into African small-scale farms and environments. It discusses aspects covering:

- Genetically modified organisms and issues of risk assessment;
- Contributions towards improving the design of monitoring and regulatory schemes basing on an interdisciplinary approach;
- Departure from European biosafety research by focusing on small-scale agriculture in Africa and help to further administrative competencies;
- Implications of their cultivation which are expected to be of importance for informed decision-making in Ghana and other comparable countries.

7.2 Genetically modified organisms in agriculture and issues of risk assessment

On a global scale, the past two decades saw an unprecedented development in the techniques of genetic modification of maize, with associated optimism about the benefits to be achieved from the development these plants. Alongside the technological package, improved weed management and a reduction in costs and amount of herbicide applications are some of the advertised benefits from these novel crops (Henry et. al., 2003). However, recent literature on gene flow suggests that for farmers, possible impacts would relate to the development of new weeds among others. Weeds are already a cost burden on agriculture, so any transgene flow that would accentuate this cost would constitute a relevant factor (Heinemann, 2007). Some threats of the widespread adoption of crops expressing Bt insecticidal protein genes include the evolution of insect resistance as a response to the strong selection pressure that will be posed on the insect populations have also been discussed (e.g. see Bardocz and Pusztai, 2007). Thus, the topic of GMOs is highly contended, and increasingly, reported uncertainties surrounding their cultivation are being made (see e.g. African Center for Biosafety, 2005; Fagan, 2007, Greenpeace, 2008). These relate to their uncertain economic value to growers, potential socio-cultural impacts and more importantly whether they are inherently safe for consumers, and the effects which they may have on the environment are largely unknown.

The coming into force of the Cartagena Protocol on Biosafety in 2000 formed an important milestone in addressing potential risk issues arising from their introduction. According to the protocol, the national admission of GMO is required to be based on a prior informed consent. For an admission, a complete risk analysis is required. If the variety was developed in the African country, the effort has to be made completely in the country itself. If it is a foreign variety for which an applicant seeks consent, the risk analysis can be based on previous risk studies and has to be completed by those information which are not covered with respect to the specificity of the country for which consent is sought. Risk assessment therefore examines extensively several cause-effect chains on a case-by-case theoretical and practical basis any direct, indirect, immediate, delayed or cumulative, undesirable or detrimental effects of GMOs on biodiversity, environment and human health. Should an irreversible, undesirable or a detrimental effect be imminent, then the decision is made whether the effect is relevant following certain scientific and legal procedures. If the effect is recognized to be negligible, then it is considered irrelevant and subsequently neglected. On the other hand, if it is relevant or significant, then the assessment of risk is taken to an extent where the remaining risks are considered to be negligible (termed residual risks). However, it is important to note that risks that are irrelevant in one environment may not be irrelevant in another environment. Thus, the need for case-specific studies on the issues of GMO risk analysis becomes increasingly important.

For a minimization of the uncertainties, it becomes important to compare available data for different situations due to differences in site structure and receiving agricultural scales and environments. The fact that international laws allow for transgenes and transgenic organisms to be protected by intellectual property (IP) (Heinemann, 2007) particularly puts small farmers in Africa at risk. The situation is particularly worrying since small farmers who chose not to grow GM crops may even be held liable and would incur liabilities in an event of transgene flow into their farms from neighbouring farms grown with GM. Potential impacts on genetic resources on large scales will have to be expected, alongside loss of valuable commercial options for farmers. Other possible consequences could be that organic food production as a potentially viable development option may be closed, since harvests are required to be free from GM impurities. In Africa, since small farms are crucial in terms of number of people fed per hectare of land (Planetdiversity, 2008), any additional losses would constitute a significant threat. Thus, this study makes a first contribution towards identifying the spatial agricultural structure and other underlying environmental and socioeconomic aspects relevant within the African context for making safe decisions.

Participation in the EU 6th Framework Project SIGMEA provided additional advantage that enabled us to test specific risk estimation parameters such as cropping distances, frequencies and spatial features under small-scale conditions for the first time. The applied protocols had been developed and empirically used internationally for large-scale farming situations in Europe. In general, risk assessment of GMOs is based on short term studies, posing a difficulty to adequately characterise transgenic plants in relation to all environmental conditions. Thus, this study demonstrated how different methodologies could be applied to make a contribution that gives a first coherent overview. This work therefore constitutes a first step towards developing West African country specific risk assessment procedures in relation to the setting up of pre-commercialization protocols in Africa. The SIGMEA data collected from the surrounding of Bremen (Germany) and Accra typify the nature of agricultural practices in Europe and Africa relating to obvious differences in their agricultural scales.

While this thesis argues that the introduction of GMOs into both agro-systems would pose a threat to the security of conventional agriculture or organic farming, there is a greater limitation towards risk management, post-commercialization testing and maintenance of safety practices on a small spatial scale in Africa as evidenced through the Ghanaian situation. For example within Germany, GenTG (The German Genetic Engineering Act) regulates the cultivation of GMOs through the measures of ex-ante regulation and ex-post liability rules (Grosse et al., 2008). Ex-ante regulation stipulates the registration by a GM farmer three months prior to planting alongside keeping all relevant safety measures which are kept on a public register. With ex-post liability, even if the farmer follows rules of ex-ante regulation, he can still be held liable in case of unexpected adversities (Grosse et al., 2008). From the perspective of the spatial agro-structure and economic conditions of local farmers in Ghana, risk regulation would implicitly constitute additional costs of GM farming. Other unbearable additional sources of costs would be damage and liability costs e.g. not get involved in court processes (see Network of Concerned Farmers, 2004).

7.3 The spatial pattern of maize cultivation practices and implications for gene flow

7.3.1 The methodological approach

Remote sensing data for assigning map positions of agro-ecological and geophysical parameters within the study area is lacking. Hence, spectral signatures of maize fields and other surfaces are largely unknown. Therefore, it was important to obtain information on the locations of maize farms and feral stands from ground-based sources through the utility of a hand-held GPS. This study thus explored exemplarily, a methodology which assessed these factors in relation to the dispersal pattern of maize as described by the mean farm size and the mean minimum distance between maize farming locations. In the context of biosafety, this is of high interest since maize locations represent potential sources of new genes to neighbouring conventional fields and related isogenic feral species. GPS as the basic survey unit, offered low cost of setting up the project, time saving at the survey sites and provided accuracy within 10 m of measurements.

7.3.2 Assessment of maize field neighbourhood distances

Pollen-mediated gene flow is influenced by spatial proximity, i.e. where for instance a GM crop and its sexually-compatible relatives occur within their respective pollination distances (Heinemann, 2007). Therefore, gene flow may occur due to the transfer of transgenes to non-GM crop or their feral counterparts. In relation to maize, the outcrossing rate between crops depends on several other factors apart from separation distance. Some additional factors such as local barriers to pollen movement (e.g. hedges, woods), local climatic conditions and topography are contributing factors (Debeljak et al., 2007). However, distance is a very important parameter that can be used to estimate how far maize pollen is transported to its neighbours. Under distance factors, Heinemann (2007) argues that there is a greater chance of gene flow if potential pollen recipients are adapted to, or grow in the environment in which the GM crop is grown. In this study, we found out that nearly 90% of all farms totalling 1,390 lay within a mean nearest distance of 100 meters. Thus, it can be concluded that the dispersal resulting from a potential cultivation of GMOs in Accra presents risks especially when actual transgenes are involved.

There were an estimated average of 3 fields within the first 20 meters from a field location, increasing to a maximum of 210 fields at a distance interval from 3900 to 4000 meters. The latter estimation has implications for gene flow with respect to wind effects and insect pollination activities e.g. by honey bees which have activity ranges of about 5000 meters. This further underscores why gene flow through pollen outcrossing at spatial scales has received utmost attention in species that are fully or partially cross-pollinated, e.g. maize in Mexico (Quist and Chapela, 2001). Studies on pollen outcrossing with distance showed that for conventional maize, 98% of pollen is deposited within 25 meters of the emitter field, and nearly 100% within 100 meters. In addition, 99% of the cross-pollination that occurs outside the emitter field takes place within 18-20 meters of the emitter field borders (Brookes et al. 2004). These suggest that for the African conditions, segregation of traits would simply not be feasible. Prevailing cropping distances means GM would not be controllable and retrievable and organic agriculture as a development path closed.

7.3.3 Assessment of feral distances from flowering field neighbours

The establishment of maize ferals in the study area may be described as being the result of seed or propagule dispersion providing a typical example of seed-mediated gene flow in this instance. However, our estimation of pollen transfer distances of maize ferals from their flowering field counterparts was done to assess the extent of potential gene exchange between farms and feral populations, or to natural near-habitats. The data showed that about 90% of all ferals occur within 85m from field crops. This presupposes that even though ferals are widespread on the landscape (Fig. 6.1.2), a majority occurs in close proximity to crop fields (Fig. 6.1.4). For instance 10%, 17%, 29%, 4%, and 3% occur within distances of 5m, 10m, 15m, 35m, and 80m from field crops respectively. Thus, indicating that a majority of feral stands occur within crop pollination range.

A reason could be that suggested by Newstrom et al. 2003 for the observed distribution pattern of ferals. They confirmed that in most cases, seeds and propagules disperse close to the source field and only a small amount of seeds spread over longer distances. In Fig. 6.1.3, it was shown that not all fields occur in close proximity to feral populations with the maximum nearest distance estimated to be 1,525 m. This may be attributed to stochastic effects i.e. occasional long distance seed dispersal, without due regard to the extent of geographical separation (see Heinemann 2007, p.17). In such situations, it is expected however that gene flow will not necessarily always decline exponentially, the farther the genetically engineered crop. The findings therefore substantially increase the challenges of the scope of risk assessment in small farming areas as evidenced in Accra.

7.3.4 Analysis of field acreage and spatial distribution

Spatial analysis is a vital part of GIS. With the help of a GPS acquired data, it was possible through a GIS operation to visualize and relate the overall agricultural structure. Display of fields was possible on Google map with an acquired resolution of about 1 x 1 kilometer (Fig. 6.1.5). Furthermore, detailed displays were feasible using ArC GIS (Figs. 6.1.9-6.1.15). The requirement to integrate field locations obtained as a ground-based GPS data into a GIS system was useful in calculating the acreages of farms cultivated by local people. The overall maize cultivation situation could be described as irregularly-structured and highly dynamic. Most cultivated plots (ca. 98%) are typically less than 0.5 ha in size (See Fig. 6.1.17 for the details). A very low number of fields (0.22% of all fields) were found to be larger than 2 ha. This result confirms that maize cultivation is typically a small-scale practice and a suggestion of its role played as the predominant mode of traditional farming in Ghana. However, under the situation where about 99% of fields occur at isolation distances of between 5-150 m pose immense risk with respect to pollen outcrossing frequency with distance (see Fig. 6.1.1).

Therefore, size distribution appears to be a crucial factor to help assess potential of gene transfer from fields to near natural habitats around the field. The length of the contact area between a field and surrounding vegetation depends not only on the size but also on the shape of the donor field. In Figure 7.1, it is shown that in some cases, a high number of smaller fields occur in close proximity to a larger field. In this situation, it is also seen that some adjacent fields are apparently joined by about 1 meter land segment. The larger number of smaller fields also provides indication of the cultivation of different seed varieties of maize including the growing of GM in the future. Gene flow is generally expected to be higher from the larger fields into the neighbouring smaller fields due to their smaller sizes. However, some dense situations of smaller fields were revealed as shown in Fig. 7.2. The data suggests that smaller farms possess longer field edges with other smaller neighbours hence facilitating the incidence of gene flow, and potentially with neighbours of interbreeding feral partners. Therefore, regions with higher cropping densities would have higher probability of accumulating transgenic genotypes in crop populations. High cropping densities suggest a greater number of plants leading to a greater amount of pollen being produced, increasing the likelihood of successful fertilization, and the likelihood that long-distance gene-flow may occur (Heidemann, 2007).

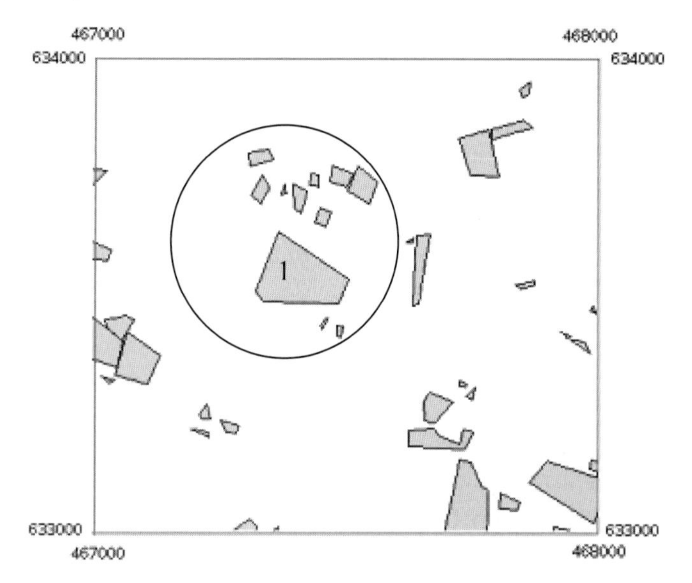

Figure 7.1: An example of the spatial configuration of a large maize field (1) in close proximity to a larger number of smaller fields. Coordinate values are in meters.

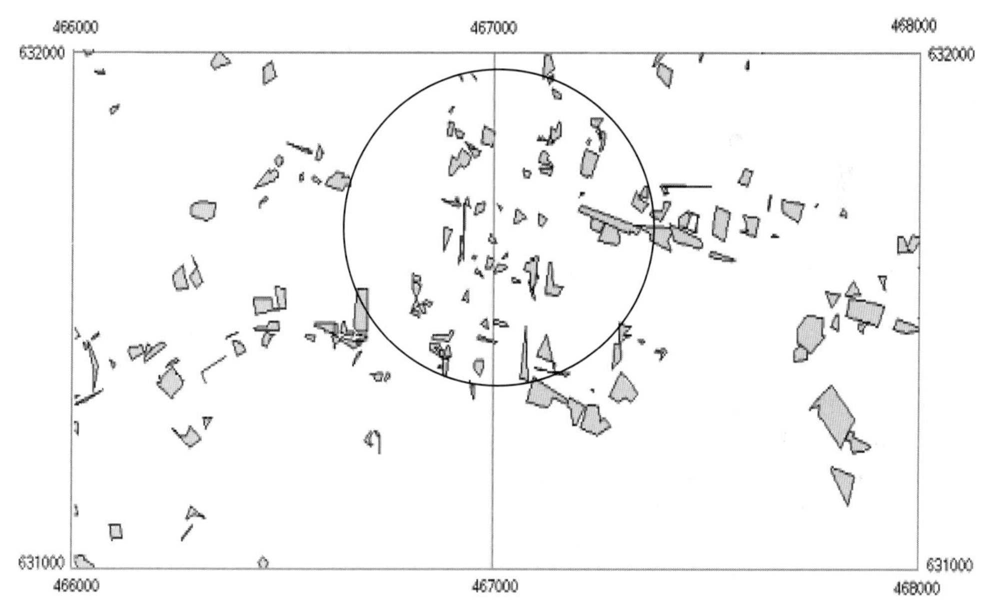

Figure 7.2: An example of the spatial configuration of maize fields depicting a dense cultivation practice. Distance values are in meters. Areas with a high density of GM fields occurring in association and sometimes contiguous with non-GM fields are likely to experience a higher cross-pollination.

7.3.5 Number of farms within certain distances and implications for monitoring over larger areas

In recent years, several researches have been conducted to estimate the applicability of separation distances as a means to control or minimize gene flow between GM and non-GM field cultivation practices over larger areas (see e.g. Henry et al., 2003; Ma et al., 2004; Messeguer et al. 2006; Weber et al., 2007). For a practical implementation of separation distance measures, it could be discussed with farmers how many farms should be allowed within specific range of distances, or the coordination of crop rotation allowing the planting of different crops in locations where maize is mostly cultivated. This study sought to investigate these features by analyzing how many field neighbours could be found within certain range of distances. Our study covered an area of 25 km^2. In order to derive spatial variability estimations of fields over larger areas, an area of 144 km^2 was analyzed by extrapolation using a mathematical algorithm that allows to cover eight adjacent areas (Fig. 6.1.6).

The findings showed a generally dense number of neighbours within the investigated distances e.g. for distances of 100 m, and 200 m about 70%, 20% of all fields have a maximum of 4 and 14 neighbours respectively that have to be expected (Fig. 6.1.7). These may be due to the small nature of fields. Palaudelmàs et al., in 2007 confirmed that the strategy of separation distances would prove difficult in areas where the size of cultivated fields is relatively small. Thus, the clustered nature of farms in Accra makes an application of separation distances a highly improbable scenario under the subsistence agriculture in Ghana. Under these conditions, the best GM confinement strategy would be to implement different planting times by farmers as a means to reduce the overlap time in flowering between donors and receptors in order to reduce the percentage cross-pollination between fields. This does not in itself appear feasible because small farms are normally rain-fed, and the timing of rains is a decisive factor determining when maize is sown by local people.

7.4 Crop population demography and implications for gene flow

7.4.1 Crop growing areas

A crucial need in risk assessment research is to identify potential hazards and relate it to the likelihood of the presence of a transgene. Aside this, the species type and the ecological setting in which transgenes or GMOs are released are important for an explicit tracking of the likelihood of risk in the environment (Fagan, 2007). We had classified the various land use sites that could be used to assess the severity of potential ecological effects on specific cultivation environments in the region. Thus, the distribution pattern of maize crops was presented in detail by differentiating the maize crop growing areas based on the differences in the urban environment or land use types (see section 6.1.8). The analysis distinguished publicly accessible parts from areas of restricted access such as private and industrial property. In the latter, permission was sought for entry. Additional land use categories classified covered a range of areas including construction areas, and small business places alongside areas of marginal usage. In all, 9 different maize land use types were recorded as shown in Table 6.5, with nearly 40% of all fields grown within home compounds having a mean field size of 742.5 m^2. Even though privately owned open-spaces constitute only 4.2% of the total number of locations, they appear to contribute more to the overall maize acreage with an average size of 3,036.8 m^2. However, home compounds could have a maximum field size of 39,590 m^2 representing the highest among all the land use types.

Traffic regions make up 29% have average acreage of 730 m^2. Construction sites constitute the third most important place for maize growing amounting to about 11%. Industrial sites accounted for the least cultivated location estimated to be about 1%. The result showed that the overall maize cultivation plots or locations occur within heterogeneous sites. Specific biosafety management and regulation need to be adequately informed on the whole. Strategies must relate these factors to the spatial conditions which indicated that even though heterogeneous, growing fields and plots are clustered or occur in very close patches to one another. Therefore, the unconfined release of GMOs into a particular site would lead to an infringement of their desired borders, inadvertently leading to transgene flow into other close-by environments and agrosystems where the GMOs were not originally introduced.

7.4.2 Flowering synchrony of farms

Male flowering analysis showed considerable overlap between fields (See Table 6.6). This may have been due to close sowing dates as well as attributable to prevailing climatic and landscape regimes (e.g. altitude, rainfall and edaphic factors). Since maize is wind-pollinated, this implies that field-to-field pollen exchange between neighbours is very probable. Kuparinen in 2006 showed that in many cases, turbulent wind flow could aid pollen movement across very long distances. These estimations were achieved though aerodynamic simulation studies. Also, other studies have indicated that pollen remains viable in high amounts in several hundred meters of heights, investigated via experiments using an aeroplane (Brunet et al., 2008). These arguments in relation to the available data heighten the uncertainties of the impact of dispersal of artificially modified genes to conventional fields by air-borne pollen. Thus, under the investigated tropical conditions, relevant flowering heterogeneity is observed which could be explained by the longer time span in sowing.

7.5 Quantifying the spatial spread of transgenes (modeling cross-pollination)

7.5.1 Quantitative models

In order to manage adventitious presence in non-GM harvest, cross-pollination needs to be quantified (Sabellek et al., 2007). In principle, the level of gene flow from GM populations into conventional crops or vice versa as depicted in Fig. 7.3 cannot easily be investigated empirically. Pollen and seeds can disperse over very long distances that their whole dispersal patterns cannot be directly observed (Kuparinen, 2006, Katul et al., 2005). Secondly, since the release of GM plants are strictly regulated, it would not be possible in practice to conduct field trials on large spatial scales to determine which information would be required for decision-making (Schiemann, 2003). Therefore, quantitative models serve as the best predictive option to estimate pollen movement in the atmosphere and outcrossing potential in order to help evaluate potential risks as a result of gene flow processes (Aylor, 2002; Breckling and Menzel, 2004; Kuparinen, 2006; Bock et al., 2002).

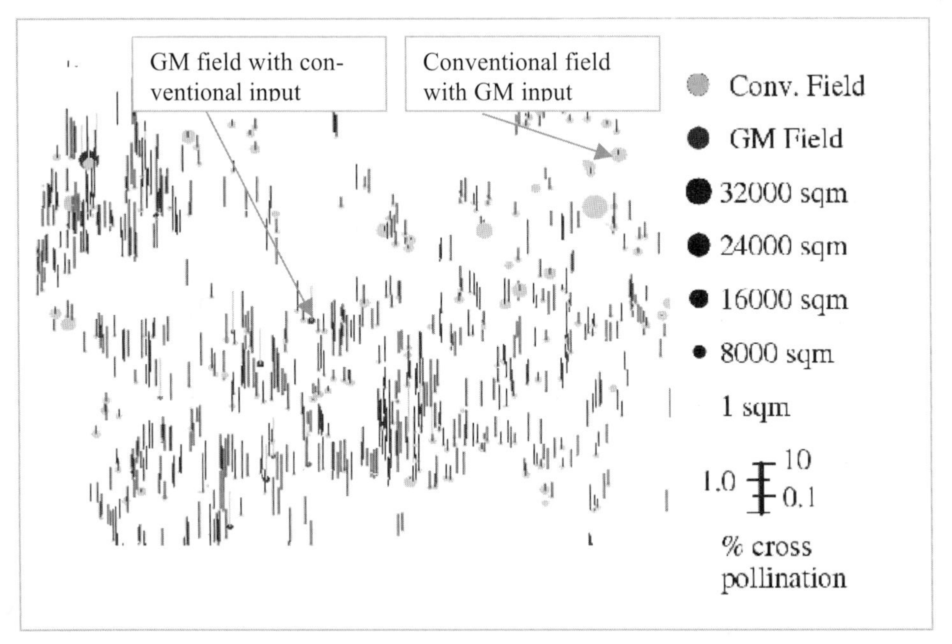

Figure 7.3: Simulation of gene flow into GM or conventional fields based on external pollen received from each field in the region aggregated for each recipient field. Note that the introgression levels are presented on the logarithmic scale with the data indicating various average contamination levels from 0-10. The results generally show that some fields do not receive external gene input. However, a larger number of smaller fields receive very high external gene inputs compared to their larger counterparts.

For the local conditions, predictions using the model (Mamo) related the distance between the donor and recipient fields, which appear to play the dominant role, field sizes, maize traits and their vegetative period. The results indicate high GM cross-pollination rates of between 0.5- 5 % regardless which seed entry scenario is applied. Even with a single GM source field among a population of 1,389 conventional fields, increment rate of GM into the overall maize area amounting to 0.21% would have to be expected. This means that transgenes would spread rapidly through the area under the set of considered attributes. Thus, on a regional scale, pollen would be shed during longer time span than is usually the case in temperate climate, and a higher level of variability in the cross-pollination is apparent in the model result calculated for Accra.

7.5.2 Simulation of regional cross-pollination through ecological modeling (The case of the Maize Model, MaMo)

The model helps to predict the average cross-pollination conditions on the regional level that is relevant for GMO manageability for a given year for given fields within the region. The model was based on several world-wide studies capturing the variability in climate and environmental factors and therefore the best available option representing the state of the art and most reasonable. It was parameterised for location, size, distance between fields, and vegetative period. However, it is important to note that specific meteorological conditions and certain environmental variables would be important to be included in the next steps of the model development to attain a higher spatio-temporal resolution of the results. The model results are to be interpreted only as regional trends and do not provide specific predictions for particular fields.

Despite this limitation, the model has been able to provide first results, as well as reasonable probability estimates based on distribution of pollination rates among the small farms. This therefore makes a contribution to regulating GMO products in the region. Depending on the prevailing conditions, field sizes up to 0.5 of a hectare could receive pollen from surrounding donor fields in a range of 0.01- ~100%. In relation to the distance, it could be argued that since on the average nearly 99% of all fields lie within a distance of 150 meters makes the above stipulated assumption supported. In attempting to directly quantify the amount of pollen contribution, I refer to a study by Hofmann in 2007 in which he empirically investigated pollen deposition (Fig. 7.4) using pollen samplers. This comparison is relevant to avoid potential limitations of evidence based on theoretical models.

7.6 Assessing the socioeconomic background of farmers and implications of gene flow

The management of GMOs presents enormous challenges for small farmers in developing countries (Heinneman, 2007). This is why some countries especially in the Africa region have adopted stringent regulatory options including GM-free decisions or moratorium in order to minimize unexpected developments emerging from their introduction

(African Center for Biosafety, 2005). According to Egziabher (2007, p.404), "In developing countries, socioeconomic considerations should have a very high priority in decision making on whether to import a given GMO or not, especially when the GMO is a commodity". In spite of the obvious need, few studies exist that review the socioeconomic conditions as a first step to interpret potential implications of GMO releases into specific African environments on a specific scale. The situation is exacerbated by the fact that while advertisements on the prospects of GM technology advances through large-scale projects, a review of literature reveal a less focus on biosafety research looking specifically into the socioeconomic conditions and livelihood implications especially on women (Meyer, 2005).

In order to assure the beneficial use and prevent unintended effects consistent with their international obligations, socioeconomic considerations have been established by the Cartagena Protocol as a key component of regulation and notification procedures of genetically modified organisms. Local participation, access to information, and information sharing are acknowledged to be major drivers for reducing risks that go beyond the efforts of single individuals and societies. This section addresses the socioeconomic context of the populace based on findings obtained from 200 interviews. This section addresses issues covering:

- General demographic and livelihood conditions;
- The management nature of subsistence and commercial farming approaches;
- Aspects of local seed acquisition, procurement and technology access;
- Issues of land ownership rights and gender aspects.

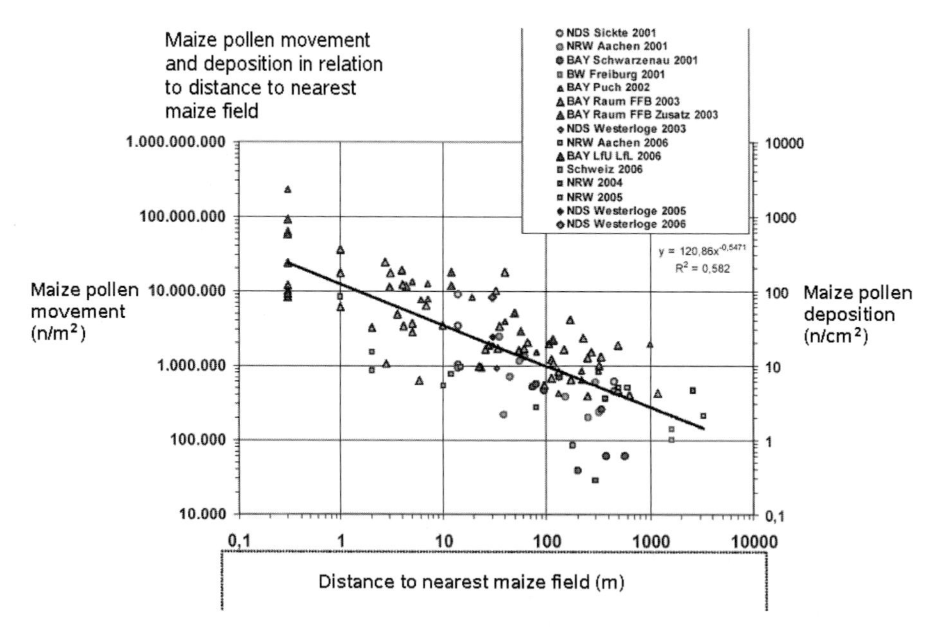

Figure 7.4: Hofmann (2007) Pollen movement and deposition in relation to nearest distance to maize field determined with a pollen sampler PMF from 2001-2006 at different locations in Germany.

7.6.1 Demographic characterization of the local farmer population

Demographic studies are important to understand the population structure of a place at a particular time. Maize growing in Accra may be characterized as a low investment activity because the cultivation is often carried out as a small-scale family enterprise mainly for use in households. Despite the low investment scale, it has the potential to generate some small income in support of the overall family budget when transacted. The results confirmed maize growing to be largely an informal sector occupation with about 80% of the sample population working in the sector. More than half (50%) of the population constitute the active age group of between 25-35 years, majority of whom are men. The data also confirmed a significant male-dominated settler population mostly working in the informal sector as care-takers of lands which they cultivate. Women respondents accounted for 35% of the sample population of farmers.

However, this does not directly suggest that the women were generally fewer because some women were remotely observed working on the farms of their husbands. Even though the actual number of women was concealed, it can be described that agricultural labour and household food production is highly supported by women. Education may be described as being limited or virtually non-existent. The data revealed a generally low educational status among the population especially women, which may have accounted for their inability to find working places in the formal sector. Among those who never had education, nearly 69% comprised of women. Their low access to education presents inherent consequences on the effective management of small family farms. Migrant population forms the bulk of farmers who do not have a formal means of living.

In terms of household size, a wide variation in the number of members was recorded. On an average, more than half (50%) of the total population of respondents had about 4 persons or more per household (Fig. 6.4.3a). However, this does not directly translate to the number of available farm labour capacity (Fig. 6.4.3b). Females have larger households compared to their male counterparts, but generally lower private incomes compared to the men (Table 6.11). The differences were calculated not statistically significant. However, female farmers tend to have a rather higher overall household income. This may be due to additional financial contribution made by their husbands and other members of households to the overall household family income. Pearsson correlation analysis showed that household income significantly relate to available labour at the household level (Table 6.10).

7.6.2 Seed acquisition criteria and technology access

The findings show that within the urban context, farmers grow seeds from a wide variety of sources, but largely obtained from quantities bought for food (Table 6.13) with the advantage that they save on cost of acquiring new seeds from formal certified sources. The observed situation is different to what was reported for rural areas where farmers mainly use seeds from the previous harvest for planting (IDRC, 2008). Among farmers set of criteria for seed selection, seeds are sown from previous harvests if they are perceived to be of high yielding or early maturing quality (Fig. 6.4.5). In this context, seed exchange among neighbours or even received as gifts and acquisition from extension services appear to be among the least preferred sources (Table 6.13). Despite the wide range of acquisition sources, on the whole, we recorded extensive usage of seed landraces in comparison with commercially-certified seeds (Figs. 6.4.6 and 6.4.7). The former often utilized to secure subsistence while the latter mostly procured for market-oriented reasons (Table 6.16). Farm households mostly growing commercial varieties were significantly wealthier probably because of significantly higher labour capacity (i.e. the number of persons engaged on-farm, Table 6.16).

The use of hybrid seeds is widespread among the population. A greater number of farmers interviewed (68%) re-sowed fractions of harvests due mainly to financial reasons. Other factors covered considerations of maintenance of specific genetic trait, early maturing or if the harvested crop was perceived to be insect resistant or of high yielding varieties. An estimated 32% of farmers did not re-sow harvested quantities due largely to easy accessibility to grain markets or certified seed shops in urban centers. The use of landraces aside the need to conserve genetic properties was largely attributed to the issue of low financial resources on the part of farmers (Table 6.14). This suggests that some farmers could not purchase improved certified varieties even if desired due to financial constraints. It is concluded therefore that regardless of the perceived inherent qualities of a technology, farmers would probably not be willing to further invest in additional factors of production that they do not possess or presents some difficulties to access. Those with no education grew more traditional varieties and vice-versa (Table 6.16).

It is worthy to note that even though farming system is limited by problems of insect pest management on-farm, insect sensitivities are least considered among farmers set criteria in terms of seed quality preference (Fig. 6.13). Thus, the adoption of insect-resistant GM crops by smallholders is potentially less probable under the circumstance of seed purchase. Medium to relatively large-scale farmers would rely on early maturing varieties to be able to sell fresh cobs or high yielding to optimize crop output and income as the data has revealed. Therefore, specific trait selection of early maturing or high yielding varieties is of key importance for this group of farmers. In some cases, a clear distinction between traits of high quality or low reproducibility was difficult to classify by farmers. Therefore, on the whole, the agricultural system presents complex structure with the implication that makes the assessment of crop purity a daunting task.

Due to the dynamic social context and proximity to metropolitan areas, it was investigated to what extent certain imported agricultural technologies such as herbicides or fertilizers were applied within the peri-urban context. It was revealed that hardly any forms of herbicides or fertilizers were used by the local people (Fig. 6.4.4). The findings indicate phenomenally low use probably due to the small nature of fields and readily available weed control measures such as hand tools, and easy accessibility to fertility options in the form of organic weed materials or manure. Aside the cost considerations, this observation could also be attributed to the small nature of fields in the magnitude of a few acres. Thus, under the present conditions, the frequent use of traditional tools and methods potentially renders the cultivation of herbicide-tolerant GM crops and associated technological packages minimally attractive to local farmers.

7.6.3 Features of the subsistence and commercial farming contexts

Commercially-oriented farmers use improved seed varieties often obtained from certified sources as a means to ensure high productivity under conditions of minimal or no usage of herbicides. On the other hand, considerable amounts of seeds sown within the subsistence contexts came from local landrace sources (Fig. 6.4.6) in many cases, saved from previous harvests (Table 6.13). This is due in part, to the low financial status of the local population (Table 6.12) and the inability of farmers to procure commercial seeds even if they consider them to be very relevant e.g. as high yielding (see section 6.4.4). Both subsistence and commercial farming approaches correlated largely as informal sector activities and could be understood as part of the usual routine (Table 6.10). The analysis showed that:

- About 93.2% of people in the informal sector grew maize for subsistence purposes;
- The remaining 6.8% in the informal sector grew maize for the market as a way to supplement family income;
- All persons (100%) working in the formal sector grew maize exclusively for subsistence purposes meaning no formal sector worker grew maize for commercial reasons.

It was also revealed that women subsistence farmers with no education significantly grew more traditional varieties (Fig. 6.4.7). The data showed that occupational factors and farm orientation (whether for commercial or subsistence) predicted agricultural land use or places where maize is grown. Commercial farmers have about two fields on average while subsistence farmers owned mainly single fields and two at the most. This finding was observed to be statistically significant. For both farming systems, crop rotation involves the growing of maize twice a year often in association with other crops such as cassava, legumes or vegetables in the major and minor seasons. The traditional usage of leguminous crops often intercropped with maize contributes to soil fertility and could partly be the reason for the low usage of inorganic fertilizers. Maize monocropping is generally not favoured and grown once per year on condition that the land is left fallow afterwards (Table 6.17) probably to allow for soil fertility rejuvenation.

7.6.4 Analysis of land ownership rights and gender factors

In the context of GMO monitoring, an assessment of farmer land rights could help to understand factors dictating where maize are grown in the communities, to reflect on space requirements, and cropping intensity factors. We refer to land rights as being the rights of ownership for persons to use land for cropping, to make improvements, as well as exclude others at community level. It allows to specify terms on which land is held, used and transacted allowing to enforce administrative and legal provisions (Adams et. al, 1999).

Three (3) types of land ownership were identified among farmers, namely private, leasehold and public land holding. In the peri-urban sector, they occur within open-spaces or homesteads either as front or backyard gardens and explain the situation why farming in marginal places is important for subsistence. Open-space farming occurs as small fields for instance along roads, with farmers holding on to such marginal areas to relatively larger areas within the community regardless whether any formal land rights exist. In open-space farming, tenure agreements are largely informal and the extent of farming on urban spaces vary often extending to wetlands, even at areas of construction or on industrial sites, just in hope of a harvest.

In general, a majority of open space farmers do not own land they cultivate with the implication that they hardly pay any rents on such lands. Private lands are mostly cultivated by care-takers, mainly settlers or migrant workers coming from diverse backgrounds, and those who 'actually' own land achieved that on leasehold basis. The municipal authorities or private developers largely own most of these lands. Examination from other studies which assessed migration-environment nexus concluded that changes in the environment due to agricultural land use intensification have mainly occurred due to population increases at destination areas, largely precipitated by in-migration of people (Codjoe, 2006; Myers, 1997; IUCN, 2000). Tenure conditions seem quite insecure and may be the reason why farmers are not motivated to invest in on-farm infrastructure, contribute to long-term fertility improvements or apply soil conservation measures.

Most women in the sample population are landless, growing maize on publicly available lands. The results showed that women with virtually no education grew maize on public lands while men do so either on privately-owned land or on leasehold basis. Migrant population is high accounting for nearly 70% of farmers who mainly grow on marginally available lands or in private areas where they are employed as care-takers. It was found out that agricultural intensification on own lands is rather limiting. Codjoe (2006) argued that migrants tend to be more aggressive in their farming practices compared to indigenous populations mainly because of insecurity of tenure.

In a number of instances, married men hold most of land with an estimated median value of 99,661 m^2, followed by the married women farmers having an estimated 39,869 m^2. For the unmarried cases, the men again hold more land in the value of 19,999 m^2 compared to the women having 6,928 m^2 (Table 6.20). These underscore the disadvantaged place of the women when it comes to areas available for cultivation in the region.

7.6.5 Socio-economic implications of gene flow

Farmers may select, manage and save seeds each growing season according to their socioeconomic and cultural circumstances. This form of plant breeding enables farmers not only to adapt seeds to their purpose but also help to expand upon local crop diversity (Heinemann, 2007). Small-scale traditional farming in Accra comprised small households often owning small plots which they cultivate and tailored to meet their own required conditions with hardly any external inputs applied. The context-specific agricultural knowledge held by farmers, associated land rights, seed exchange practices, as well as crop rotation practices is very important to sustainable agriculture in developing countries. However, seed exchange is a form of human-assisted gene flow potentially enhancing the risks of dispersing transgenic traits to other fields or environments and increasing the potential for gene flow through fertilization (Heinemann, 2007). The context of seed acquisition and exchange practices in Accra involves a complex process of interactions among farmers with a consequence that seed admixing is very likely making trait segregation a very challenging task.

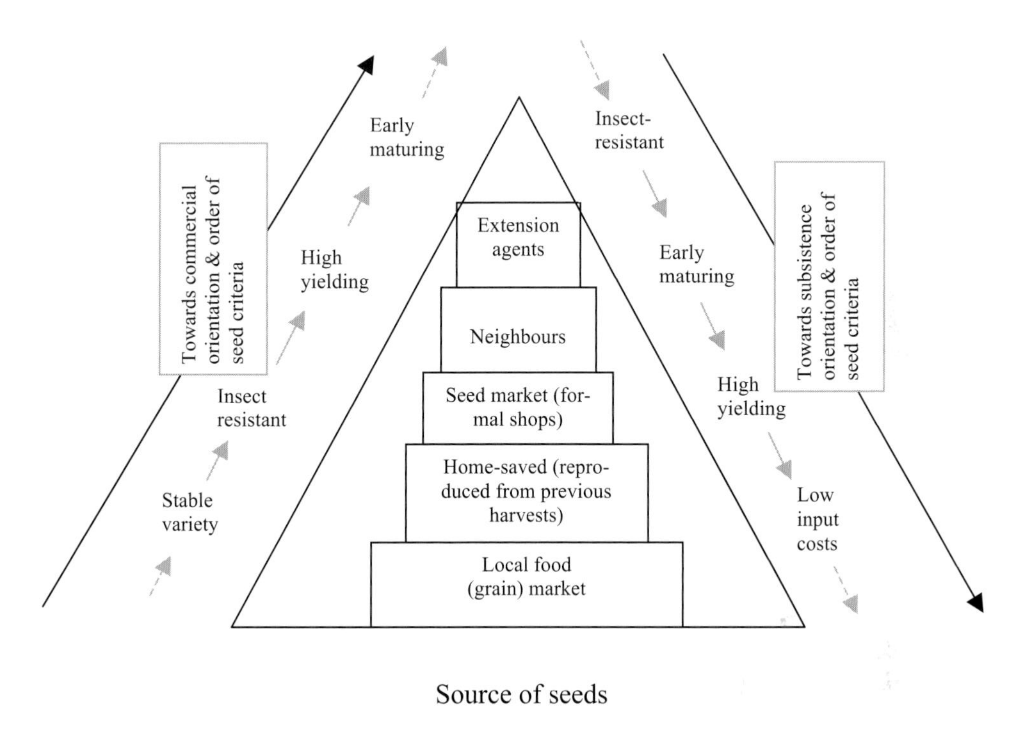

Figure 7.5: Farm orientation and levels of seed selection criteria in Accra West in 2006. In most cases, improved seeds reach farmers through the 'formal' sector, namely small shop enterprises and extension agencies where farmers buy commercially certified seeds. A majority of the farmers acquire the seeds through the 'informal' sector, mostly grain markets (for food), local level reproduction from previous harvests. Subsistence farmers acquire improved varieties from shops with a perceived notion to sell part of their harvests.

Again, seed exchange practices has the potential to limit the prospects of coexistence. Coexistence is expected to enhance farmers ability to provide consumers a choice between conventional, GM and organic products. This has the implication to lower the economic value for conventional or organic food certifiers. If this happens in the commercial sector, those who may end up growing transgenic crops either by accident or intentionally could be affected by export market rejection or risk losing market access as worse implications. Livelihoods may be compromised due to the decline in crop genetic purity and loss of farmer trait preferences. The subsistence sector is characterized by traditional agriculture, and maize growing sometimes intercropped with other staples and the application of crop rotation contributes to overall family income. Often described as unproductive, subsistence farming can play an important economic, cultural and food security role in a society (Brüntrup and Heidhues, 2002).

Since the practice is widespread among small farmers and seed saving guarantees seed supply year after year, the unanticipated transgene flow would therefore impose serious repercussions on smallholder's freedom of choice for agricultural biodiversity and maintenance of desired crop traits. National and international laws allow for transgenes and transgenic crops to be protected as intellectual property (IP) (Rosendal et. al., 2006). Therefore, legal prosecutions may ensue against farmers especially if they fail to acknowledge the seed producer's intellectual property. This could be a serious problem for small farmers in the African region who do not intend to grow transgenic crop varieties. Patent infringement is considered as a very serious issue and as was the case of Schmeiser vrs. Monsanto in the popular case of patent infringement on the part of Schmeiser that the provisions of the Canadian Patent Act did not provide extensive IP protection for molecularly engineered gene, effectively nullifying the farmer's classic property rights in his plants or seeds (Heinemann, 2007; De Beer, 2005).

Traceability systems supposed to track through the supply chain for food safety and quality control would be very difficult to implement because of the inherent difficulty of product differentiation among small farmers. This is expected to render recall of a potential GM release almost impossible. Even if GMO labeling should be mandatory, for regulators, traceability would constitute enormous challenge and the implementation of a post-market monitoring measure or product quality assurance would be far fetched. Owing to the farming structure, it is possible that gene flow even at low frequencies may be detected in the crops of farmers of conventional or organic production. By application, they indirectly become liable for the GM content. Where laws regulating this form of product harm are applicable claims in the order of magnitude similar to the Starlink corn recall may come up. The economic impact of the Starlink corn recall was estimated at US$ 1 billion to Aventis alone illustrating that the magnitude of claims could go up, especially when later detected (CBD, 2006; Greenpeace 2006). Transgene flow does not only increase the liabilities of non-GM producers but also has the potential to increase the liabilities of GM- producers who receive transgenes from other GM varieties. According to Heinemann (2007, p. 53), this may have adverse effects on societies with traditional, subsistence and seed saving cultures. African farmers who may wish to enter the international organic market may not obtain market access. Prospects are very unlikely that farms that may be affected by transgene contamination would receive organic certification for their products and those already in the organic export programme may lose approved certification. In relation to this, Heinemann confirmed in 2007 that the potential economic damage from an inability to provide certified organic or non-GM products due to transgene contamination could be significant with dire consequences. Within the EU for instance, the average price premium paid for organic products is between 20%–50%, while in Japan for example, it was revealed to vary between 100% and 200%.

7.7 Analytical procedure for maize seed variety differentiation at the molecular genomic level

7.7.1 Experimental genomic steps and limitations

In this study, attempts have been made to characterise the relations between sampled farmer landraces and commercial seed varieties on the genomic level. The aim was to establish an analytical procedure for seed variety differentiation in natural populations on the molecular level. This was thought of as additionally useful field that could open up the biosafety context in Ghana to equally address crucial issues related to seed purity and conservation genetics. However, a problematic issue was the limitation of funding resources to adequately investigate to bring out sufficient differentiation or similarities between the samples. For instance, we were unable to develop phenograms (tree-diagrams) for the main reason of weak differentiation resulting from a lack of commonly shared characters and a lack of characters at all that would allow for discrimination between subclasses or taxa within the dataset. A probable causal factor was the limited number of primers used for this investigation. We had used only four primer pairs. For a good distinction between samples more primer pair sets are required (see e.g. Senior & Heun, 1993, Warburton et al., 2002, Magorokosho, 2006). For the resulting gels of primer phi 331888, a tree construction was attempted with the help of the FreeTree programme[5].

This dataset could only discriminate 66 samples with only 11 characters (observable bands), which is a difficult relation when trying to detect differences and similarities. Thus, for a further refinement, it would have been useful to add more character sets from other RAPD or SSR primer sets to this initial data, which may have resulted in a better resolution of the resulting tree. Another problematic issue was that the seeds brought from Ghana failed to germinate under the conditions we provided at the University of Bremen, where the laboratory study was executed. Perhaps, these were sub-optimal for tropical species. Hence, a DNA extraction from seeds was the only feasible option with higher extraction rates obtained through the Chelex® method as against the DNAeasy Plant Kit. Even though Chelex® was a useful method expected to eliminate protein inhibitors possible contamination may not have been excluded resulting in the weak DNA band fragments. Lack of seed germination was partly attributed to dormancy since the seeds had been kept for rather very long time of over a year, stored under varied conditions with a chemical treatment using Actellic dust (chemical mixture of Pirimyphos-methyl and Permethrine) for protection against pests such as weevils. Possibly, the dormancy could have been broken with the use of geberellins. Unfortunately, these emerging aspects are resource intensive and could not allow for further genomic steps. Further funding would be required to proceed in this field.

---

5 FreeTree is a program for construction and bootstrap (and jackknife) analysis of phylogenetic trees from binary data (RAPD, RFLP, AFLP etc.), http://www.natur.cuni.cz/~flegr/programs/freetree.htm.

These aspects are very relevant because landraces represent a valuable genetic resource and an important cultural heritage. From the conservation-genetics point of view, it is of particular importance that they are preserved "in situ", where populations are cultivated under the continuous pressure of various evolutionary forces. This dynamic conservation of landraces, in contrast to genetically modified crops, implies that seeds are reproduced on farms year after year. This aspect is unique to landraces (Bitocchi et al., 2005). This thesis is therefore of the view that the option of easily accessible and the widespread possibility to exchange seed landraces among farmers constitutes a high economic incentive for the smallholder within the subsistence context.

7.7.2 Relevancc of molecular DNA markers in gene flow estimation

For further investigation steps into these issues, a review of specific methods using molecular genetic markers to estimate actual gene flow is necessary. Gene flow estimation using DNA markers involves identifying a plant in a population with a unique genetic marker (e.g. an isozyme allele) and follows the appearance of the marker in the subsequent generation. Transgenes can act as markers for tracking gene flow and results from various studies have been documented (Quist, 2004; and Holck et al., 2002). In terms of estimating genetic variability, several DNA based techniques exist, and tested for their suitability to function within high-throughput screening processes. These techniques have been investigated for their suitability to evaluate interactions between individual organisms at regional levels (Burkhardt, 2006).

The main DNA techniques present are RFLP (Restriction Fragment Length Polymorhism), RAPD (Rapid Amplified Polymorphic DNA), AFLP (Amplified Fragment Length Polymorphism) (Marsan, 1998; Jones et al. 1997) and SSR (Simple Sequence Repeats) (Gupta and Varshney, 2000). SSRs are a relatively new class of genetic markers. They are a subset of tandemly repeated DNA family represented by extremely short nucleotide sequence repeats (from 1 to 5 base pairs (bps)) that are abundantly present and interspersed in eukaryotic genomes (Taramino and Tingey, 1996). Owing to the high rate of variation in the number of repeat units, the polymorphism level shown by SSRs is high. They are easy to analyze by means of polymerase chain reaction, using flanking sequence primers (Taramino and Tingey, 1996; Warburton et al., 2002). SSR has become an ideal molecular marker in the identification and differentiation of plant varieties (Xin et al. 2005), and have been widely applied in the context of maize germplasm mapping (see Dubreuil et al. 2006; Warburton et al., 2006; Magorokosho, 2006; Taramino and Tingey, 1996; and Senior, 1993).

7.8 Potential agronomic and environmental impacts of gene flow

The environmental consequences of a genetically modified organism will depend on the characteristics of both the GM organism and the receiving environment. For example, the agronomic benefits arising from a given GM crop will be closely related to the local severity of the problem it is designed to solve. Hence, detailed information on the characteristics of both the GM and the receiving environment, including the problem that it is meant to be solved is needed to shape and guide risk assessment efficiently (Hilbeck et al. 2004). The three primary sources of GM impurity in non-GM crops are through sown seed, volunteer weeds left by a preceding GM crop, and cross pollination between crops simultaneously in flower. Secondary sources arise through the transport of seed out of GM fields (on vehicles) to cropped and uncropped land and through introgression with wild relatives.

Over time, a complex network of potential sources may accumulate. As shown from the gene flow estimations data, introgressed genes will be very difficult to eliminate within the small-scale context, with the implication that crop farms or gardens would assume inherent properties of GM plants leading to a loss of genetic diversity of local farmer varieties. The data shows that small and medium-sized farms represent efficient sources of food production when measured in terms of people fed per acre of farm land. In a scenario of potential introduction of GM maize would suggest that farmers would have no chance to compete with technologically-driven often chemical intensive system of production, otherwise not largely the case of the already practiced system of farming in Accra. Thus, genetically modified crops of corporate driven agricultural monocultures may lead to a collapse of traditional food production, and potentially huge loss of farmer livelihood at the local level.

There is relevant preoccupation and that it may not be reasonable to put people at risk who depend crucially on a specific crop without having alternatives.

Chapter 8: Conclusion

In all, 1,390 maize fields were found within the 25 km^2 study area of Accra during the study period with a mean density of 56 locations per km^2 and mean acreage of 808.7 m^2 (0.08087 hectares). Feral population are evident on the landscape amounting to 69 feral stands. About 98% of all field cases are small in size, less than half a hectare and in very close proximity with each other in the range of 5 -150 m. Highest GM cross-pollination rates into overall maize area and total number of conventional fields for seeds acquired through seeds acquired through the formal system amounting to rates of 7.7% and 6.02% respectively. The results therefore lead to the following conclusions.

8.1 Ecological and biosafety management implications

The spatial nature of the agricultural structure: The spatial configuration of the fields would facilitate the possibility of transgene flow through higher cross-pollination among small field neighbours. The existence of single larger fields with higher number of surrounding smaller fields is confirmed. This is of special interest to estimate impacts not only on smallholders but on a wider number of smaller fields with potentially diverse locally-grown seeds. Therefore, GMO seeds if introduced would pose a major challenge to maintain GMO-free zones also due to the prevailing systems of seed acquisition and local exchange which would pose a further complexity.

Field densities: The high density of fields would pose a major difficulty to practice any legal isolation distance requirements within the small-scale setting. In this context, the presence of feral maize on the landscape would contribute to the adventitious persistence of transgenes in the local maize gene pool.

Trait segregation: Distinction is hardly made between food grain and seed grain with respect to seed sown since smallholders cultivate various landrace open-pollinated seed varieties alongside available commercial hybrids. This makes trait segregation highly impossible. Thus, 'Coexistence' of GM and local non-GM systems as being tested under the European conditions would be impractical for the African situation. In theory, different planting times among neighbouring farmers as a management measure seem not practically implementable because most small farmers do first plantings to take advantage of the rains, largely not under their control. Thus, co-existence of GM and non-GM crops would be an impractical scenario under the small-scale context.

Nature and genetic resource conservation: Introduction of GMOs would lead to transgenic contamination of local seeds, with a high probability of impacting areas where organic food farming[6] is likely possible to be practiced leading to economic and

6 This is used here to mean the usual application of only naturally available fertility improvement measures employed by local farmers in crop growing and management which also largely excludes the use of chemical pesticides and inorganic mineral fertilizers. To a large extent, subsistence farming would fulfil the organic certification requirements.

socioeconomic impacts. Consequences on native biodiversity can not be overruled. Such infringement of natural borders has been described as a crucial ethical concern (see Backhaus, 2004).

Consumer choice and liability issues: The data shows that the possibility to allow for consumer choice and trait segregation is largely minimal. Thus, containment of GM products, including potential mitigation or removal from the environment if undesirable results should become apparent would require serious and realistic provisions to enforce within the Ghanaian context. Central planning of production and regulation of agricultural products marketing would be difficult to implement. To orient producers to meet the needs of consumers would constitute a daunting task since the producers are mostly the consumers of their own production in themselves.

Biosafety monitoring implementation: Biosafety measures are comparatively more difficult to implement within the African context, in relation to the developed country context, since the resources that can be invested in the establishment of anticipatory regulatory efforts or enforcement are substantially less. On a policy level, some gains have been achieved through the established National Biosafety Frameworks, which provides basic regulatory guidance regimes, as a basis to move forward on the issues. A key measure would be to effectively regulate food and feed products imported into the country since once they are admitted, it would not be feasible to control their spread in the environment should they contain GM maize or comparable GMOs.

Civil participation: Public awareness on GMO issues and discussion are limiting. Coherence in the administrative and regulatory structures is limited, as could be seen in the operation of extension services in urban areas. Dealing with uncertainties and contradiction are among key areas that need to be addressed.

Livelihood strategies and implications on women in agriculture: The data implied a high number of persons supported per acre of household farm. However, women constitute the most insecure group often cultivating on publicly free spaces. They are compelled due to their low educational status and their inability to secure jobs in formal employment. For them the introduction of GM may imply to depend on external proprietary seeds. This would pose additional uncertainty.

The issue of financing of biosafety research: A major setback is a limitation in the capacity and financing of biosafety initiatives, which affects biosafety implementation efforts and limits the effectiveness of risk assessment procedures. Independent research that is not influenced by external business interests should be encouraged.

In the context of this study, the following hypotheses were examined:

Our study investigated the hypothesis that "Cross field fertilization decreases with increasing field distance and increasing field size. Cross fertilization is higher between small and adjacent fields". The data showed higher cross-pollination rates in a larger number of smaller-sized fields adjacent to a single GM field, and therefore the hypothesis could be said to be supported. For example in Fig. 6.3.16, a seemingly hyperbolic functional relation is revealed in a 3D display relating percentage cross-pollination to the fractional area of conventional field, and distance to a single GM field. It indicates that a higher cross-pollination is obtained for most small fields of a few square meters (e.g. ~ <5000 m^2) occurring within relatively short distances (~500 m) and vice versa.

The smaller the operated plots, the higher the heterogeneity of seed sources implicitly leading to increased genetic exchange and variability. The data showed that small fields frequently occurred highly dense often in very close association with one another and a likelihood of experiencing higher cross-pollination. Such contiguously close fields in most cases belonged to different farmers with the suggestion that sown seeds were obtained or procured from widely diverse sources. This leads to a high potential of an increased genetic exchange between crops with a possible consequence of a high genetic variability within and between crop fields. This would possibly impact a large number of conventional fields under a condition of GM growing in the future.

There is significant phenological synchrony between cultivated maize crops and their wild feral populations. This hypothesis was supported due to considerable cross-pollination potential between cultivated crops and feral locations (Fig. 6.2.8). The data shows that 10%, 23%, and 28% of all feral locations occur within 5 m, 10 m, and 15 m respectively from flowering farm crops. Longest distance to a farm crop in flower was recorded at a distance range of 75-80 m. This poses implications for gene flow between crop plants and feral stands in the wild.

The use of traditional and commercial seed varieties does not differ significantly among subsistence and market-oriented farmers. This hypothesis was shown to be false (Table 6.16). Subsistence farmers mostly relied on seeds from traditional sources or landraces as a major source of sown seeds while market-oriented farmers relied largely on commercially seeds frequently obtained from formal seed shops, and vice versa. Either of the differences was found to be statistically significant.

There are significant gender differences in farm resource ownership, number of farms, and acreage, with women mostly in the informal sector growing maize mainly for subsistence. This was confirmed true on the basis of this assessment and for the fact that women are landless or hold fewer acreage of lands (median value of 218 m^2) while most of the land resources are in the hands of men (median value of 309 m^2). Men have a significantly higher number of farms compared to the women (Table 6.19).

Male farmers and their female counterparts do not significantly invest in off-farm seeds and other productivity-enhancing technologies due to generally low income status. Financial constraint is an important factor dictating seed type choice. Food grains and home-saved seeds are most relevant sources due to their perceived low input costs with respect to their mode of acquisitions (Section 6.4.4).

Therefore, this thesis sees the implementation of the "The Precautionary Principle" as a key step towards achieving potential irreversible damages due to GMO introduction within small farming contexts in the country. The precautionary principle proposes measures to anticipate, prevent and attack the causes of environmental degradation. Where there are threats of serious, irreversible damage, lack of scientific certainty should not be used as a reason for postponing measures to prevent environmental degradation". (From the Bergen Declaration, 1990, as cited by Cameron & Abouchar in 1991).

8.2 Limitations and recommendations for further research

Biosafety issues in the region suffer considerable scientific knowledge gaps. Agronomic practices are widely variable and undocumented. In this light, the following research needs are proposed:

(a) Specific forms of local adaptation: This study was characterized by short term experiments and the applied characterization may not pick up all environmental effects to determine adequately all relevant pre-commercialization testing protocols. To adequately validate the risks, long-term monitoring is proposed as a follow-up to document the trends to detect effects which were not predicted in this study. Documentations obtained over time could confirm the precision of pre-release protocols e.g. monitoring seed flow due to imports from harbours to markets. Assessing the specific influence of cropping phenology, sowing date, sown variety, would be also of main concern. Further genetic studies or diversity screening is also encouraged, towards developing an inventory of maize genetic resources for specific biosafety measures. Furthermore, the impacts of costs in monetary terms of GM crop production should be investigated.

(b) Modeling prediction: The applied model assumed homogeneous flowering phase throughout all field locations but with random deviations. The simplification limits the possibility to estimate the effects of heterogeneous flowering on cross-pollination rates both in-field and between fields. Further studies are proposed to make the model applicable to heterogeneous flowering conditions. To assess the specific influence of different factors, the existing climatic variability (wind, temperature, and altitude or slope effects) and their explicit consideration in regional models would increase the predictive value of the model.

(c) Understanding the role of barriers: Barriers pose a major influence of cross-pollination in urban areas must be further investigated. This is because physical barriers (e.g. trees, hedges) can affect pollen dispersal and cross-pollination. Impact varies according to location of barrier to receptor crop. Barriers located immediately before a receptor crop tend to reduce cross pollination levels. If a barrier comprises rows of maize between emitter (e.g. GM crop) and receptor (e.g. non GM) maize crops, these acts as a buffer, reducing levels of cross-pollination.

(d) Upscaling methods and application on other relevant crops: In relation to herbicide resistance, the canola cultivation example in Europe as has been presented in this thesis points to the fact that resistance may be transferred to wild feral relatives of the crop. This would constitute additional important aspects to be considered in risk assessment to investigate other relevant crops in Ghana for which genetic modifications have been achieved.

(e) Local peoples' participation in the GMO discussion: A large number of consumers prefer food produced from non-modified plants, and therefore the non-modified crop has an economic advantage. Thus, the breeding of modified and non-modified crops is an economic threat to farmers growing conventional plants if new genes move into the seeds. Moreover, the new genes are usually patented and the farmer can be held liable to pay license fees for sown crops (Schmeiser, 1999). For the Ghanaian situation, maize has such a critical nutritional role because it is the most important staple food crop across sub-Saharan Africa. Traditionally, it is consumed as a starchy base in a wide variety of dishes including gruels, porridge, and pastes and fed widely as paste to weaning children (Badu-Apraku et al. 2000). Thus, programmes to involve local people on aspects of seed saving, seed and crop management would improve their response towards the issues at the grassroots level.

In conclusion, this study confirms that within the small-scale context in Ghana, maize producers are largely the consumers of their own production. Therefore, within the informal sector, to orient producers to meet the needs of consumers would constitute a daunting biosafety task. A key management step would imply to effectively regulate food and feed products imported into the country regardless whether it is necessary food import or aid since once they are admitted, it would not be feasible to control their spread in the environment should they be later found to contain GMOs.

References

ABSF News (2004) Africa: Food Shortages that Need Not Be There. African Biotechnology Stakeholders Forum, Nairobi. 2nd Quarter Publication.

ACDI/VOCA (2003) Genetically Modified Food: Implications for US Food Aid Programs. Food for development Division, Washington, DC. 2nd ed. April, 2003.

ActionAid (2003) GM Crops- Going Against the Grain. May 2003. http://omg.ngo.ro/documente/gatg.pdf, accessed on 13.05.07.

Adams, M., Sibanda S., Turner, S. (1999) Land Tenure Reform and Rural Livelihoods in Southern Africa. ODI Natural Resource Perspectives. Number 39, February 1999.

African Center for Biosafety (2005) GMOs in African Agriculture - Country Status. http://www.biosafetyafrica.net/central.htm, accessed on 5.09.05.

African Draft Model Law on Safety in Biotechnology (2001). http://www.africabio.com/policies/MODEL%20LAW%20ON%20BIOSAFETY_ff.htm, accessed on 7.05.06.

Agbios [Online] Maize- ESA . MON810. http://www.agbios.com/cstudies.php?book=ESA&ev=MON810&chapter=Appdx1&lang=, accessed on 3.3.2007.

Aheto, D. W. and Breckling, B. (Submitted) Oilseed rape cultivation in Northern Germany: Cultivation density and neighbourhood distances analysed. Proceedings, Theory in Ecology – Peter Lang, Europäischer Verlag der Wissenschaften, Frankfurt).

Al-Hassan, R. and Jatoe, J. B. D. (2002) Adoption and Impact of Improved Cereal Varietie in Ghana. Paper prepared for the Workshop on the Green Revolution in Asia and its Transferability to Africa. FASID 8-10 Dec. 2002. Tokyo, Japan. http://www.fasid.or.jp/chosa/forum/fasidforum/ten/fasid10/dl/2-7-p.pdf, accessed on 7.05.07.

Ammitzbøll H., Jørgenson R. B. (2005) Transgene expression and fitness of hybrids between GM oilseed rape and Brassica rapa. Presentation at a SIGMEA WP2 Meeting in Montpellier, France on 16 November 2005 (Risø).

Andow, D. A. and Hilbeck, A. (2004) Bt Maize, Risk Assessment and the Kenya Case Study In: Hilbeck, A. and Andow, D.A. (eds) (2004) Environmental Risk Assessment of Genetically Modified Organisms Volume 1: A Case Study of Bt Maize in Kenya. CABI Publishing, Wallingford, UK.

Andow, D.A. and Hilbeck, A. (eds) (2004) Environmental Risk Assessment of Genetically Modified Organisms Volume 1: A Case Study of Bt Maize in Kenya. CABI Publishing, Wallingford, UK.

Association for Freedom of Choice and Correct Information [Online]. GMO Pharmaceutical Crops May Contaminate Food Supply- Say Scientists. December 15, 2004. http://www.laleva.org/eng/2004/12/gmo_pharmaceutical_crops_may_contaminate_food_supply_say_scientists.html, accessed on 17.05.07.

Asuming-Brempong, S. and Asafu-Adjei, K. (2000) Estimates of Food Production and Food Availability in Ghana: The Case of Year 2000. http://www.aec.msu.edu/fs2/mali_pasidma/report00/ghana00.pdf, accessed on 12.05.07.

Aylor (2002) Settling Speed of Corn (Zea mays) Pollen. Aerosol Science. 33: 1601-1607.

Aylor (2003a) Rate of dehydration of corn (Zea mays L.) pollen in the air. Journal of Experimental Botany, Vol. 54, No. 391, pp. 2307±2312, October 2003

Aylor (2003b) An Aerobiological Framework for Assessing Cross-pollination in Maize. Agricultural and Forest Meteorology. 119 (2003) 111-129.

Aylor, D. E. (2004) Survival of Maize (Zea mays) pollen exposed to the atmosphere. Agricultural and Forest Meteorology. 23 (3-4):125-133.

Aylor, D. E.; Schultes, N. P.; Shields, E. J. (2003) An Aerobiological Framework for Assessing Cross-pollination in Maize. Agricultural and Forest Meteorology. 119 (2003) 111-129.

Backhaus H. (2004) Conceptualizing modified organisms as sources of hazards and risk. Visions of science and regulation In: Breckling, B. & Verhoeven (eds): Risk Hazard Damage- Specification of Criteria to Assess Environmental Impact of Genetically Modified Organisms. Bonn (Bundesamt für Naturschutz). Naturschutz und Biologische Vielfalt 1. Pp.145-161.

Badu-Apraku, B., Twumasi-Afriyie, S., Sallah, P. Y, K., Haag, W. Asiedu, E., Marfo, K. A., Dapaah, S., & Dzah, B. D. (2000) Registration of 'Obaatanpa GH' Maize, Technical paper, CGIAR.

Balthazar & Schoper (2002) Crop to crop gene flow: dispersals of transgenes in maize: Proceedings of the 7th symposium on biosafety of GMOs, *In*: Henry C. et al (2003) Farm scale evaluations of GM crops: monitoring gene flow from GM crops to non GM equivalents in the vicinity: part one forage maize, DEFRA Report. EPG/1/5/138.

Bannert, M. (2006) Simulation of transgenic pollen dispersal by use of different grain colourmaize. Dissertation Eidgenössische Technische Hochschule Zürich. Nr. 16508.

Bardocz, S. and Pusztai, A. (2007) Post-commercial testing and monitoring (or post-release monitoring) for the effects of transgenic plants *In:* Traavik, T. and Li Ching, L. (eds.) (2007) Biosafety first- Holistic approaches to risk and uncertainty in genetic engineering and genetically modified organisms. Tapit academic press. Pp. 507-520.

Bateman, A. J. (1947) Contamination of seed crops – II. Wind pollination. Heredity 1: 235-246.

Benbrook, C. M. (2004) Genetically Engineered Crops and Pesticide Use in the United States: The First Nine Years. BioTech InfoNet. Technical Paper Number 7.

Bennet-Lartey, S. O; Boateng, S. K.; Markwei, C. M.; Asante, I. K.; Ayernor, G. S.; Anchirinah, V. M; Odamtten, G. T; Abbiw, D. K.; Ekpe, P.; Eyzaguirre, P. B. (2007) Home Garden Systems in Ghana and their Contribution to Germplasm Flows. Plant Genetic Resources (PGR) Newsletter- FAO-Bioversity. Issue No. 146. Pp. 33-38.

Benzler, A. (2004) Effects of Genetically Modified Organisms on Biodiversity and a Contribution from the viewpoint of Nature Conservation In: Breckling, B. & Verhoeven (eds): Risk Hazard Damage- Specification of Criteria to Assess Environmental Impact of Genetically Modified Organisms. Bonn (Bundesamt für Naturschutz). Naturschutz und Biologische Vielfalt 1. Pp. 13-21.

Bergen Declaration, 1990, as cited by Cameron & Abouchar. Boston College International & Comparative Law Review 14: 1-28, 1991.

BiotekAfrica News (2003) Farmer Field School for Dissemination of Tissue Banana Technology. A Topical Newsbrief. ABSF & KBIC. Nairobi. Issue 2.

Bitocchi E, Sbano G, Maggioni L, Papa R. (2005) Determining the impact of past gene introgression from hybrid varieties into landraces. Presentation at a SIGMEA WP2 meeting in Berlin, Germany on 3-4 April 2005 (UNIVPM).

Bock, A-K, Lheureux, Libeau-Dulos, M., Nilsagard, H., Rodriguez-Cerezo, E. (2002) Scenarios for Co-existence of Genetically Modified Conventional and Organic Crops in European Agriculture. Joint Research Center, European Commission.

Breckling, B. and Menzel G. (2004): Self-organised Pattern in Oilseed rape Distribution- An Issue to be considered in Risk Analysis In: Breckling, B. & Verhoeven (eds): Risk Hazard Damage- Specification of Criteria to Assess Environmental Impact of Genetically Modified Organisms. Bonn (Bundesamt für Naturschutz). Naturschutz und Biologische Vielfalt 1. Pp. 73-88.

Brookes G., Barfoot, P., Melé E., Messeguer, J., Bénétrix, F., Bloc, D., Foueillassar, X., Fabié, A., Poeydomenge, C. (2004) Genetically Modified Maize: Pollen Movement and Crop Co-existence. PG Economics. 2004, http://www.pgeconomics.co.uk/pdf/Maizepollennov2004final.pdf, accessed on 4.05.07.

Brough D. (2001) Italy Police Seize More Monsanto Seed in Raid. Reuters News Service. Reuters News Service. April, 10, 2001. http://www.mindfully.org/GE/GE2/Italy-Seizes-Monsanto.htm, accessed on 4.05.07.

Brunet, Y; Dupont, S; Delage, S; Tulet, P; Pinty, JP; Lac, C; Escobar, J. (2008) Atmospheric modelling of maize pollen dispersal at regional scale. Presentation at GMLS 2008, Bremen (www.gmls.eu).

Brüntrup, M. and and Heidhues, F. (2002) Subsistence Agriculture in Development: Its Role in the Processes of Structural Change", Discussion Paper No. 1, Institute of Agricultural Economics and Social Sciences in the Tropics and Subtropics, University of Hohenheim

Burkhardt, U. (2006) Report for the Specific Targeted Research Project under the Sixth Framework Programme of the European Community (2002-2006) for the Project 'SIGMEA' Workpackage 2, Tasks 2.2 and 2.4, UFT-University of Bremen.

Burris J. S. (2003) Adventitious pollen intrusion into hybrid maize seed production fields. American Seed Trade Association.

Calvo, R., Carrion, F., Aquino, P. and Heisey, P. (1998) The World Maize Economy: Current Issues. World Maize Facts and Trends 1997/1998. CIMMYT.

Cartagena Protocol on Biosafety (2000) (under Convention on Biological Diversity, CBD) http://www.biodiv.org/biosafety/protocol.shtml, accessed on 4.05.07.

CBD (Convention on Biological Diversity) (1992). http://www.biodiv.org/convention/default.shtml, accessed on 4.05.06.
CBD (Convention on Biological Diversity) (2006) Determination of damage to the Conservation and Sustainable Use of Biological Diversity, including Case-Studies. 2nd Meeting of the Open-Ended Ad Hoc Working Group of Legal and Technical Experts on Liability and Redress in the Context of the Cartagena Protocol on Biosafety, Montreal, 20-24 February 2006.
CEC, Commision for Environmental Cooperation (2002) The effects of Transgenic Maize in Mexico. Key findings and recommendations. Maize and Biodiversity. http://www.cec.org/files/PDF//Maize-and-Biodiversity_en.pdf, accessed on 5.09.05.
Christou, P. (2002) No Credible Evidence is Presented to Support Claims that Transgenic DNA was Introduced into Traditional Maize Landraces in Oaxaca, Mexico. Transgenic Res. 11, iii-v.
City farms [Online] http://journeytoforever.org/cityfarm.html, accessed on 18.05.07.
Clark, A. (2004) GM Crops are not Containable In: Breckling, B. & Verhoeven (eds): Risk Hazard Damage- Specification of Criteria to Assess Environmental Impact of Genetically Modified Organisms. Bonn (Bundesamt für Naturschutz). Naturschutz und Biologische Vielfalt 1. Pp. 91-108.
Clive, J. (2007) Biotech Crop Countries, ISAAA (International Service for the Acquisition of Agri-biotech Applications, http://www.isaaa.org/default.asp, accessed on 11.06.2008.
Codjoe, S.N.A. (2006) Migrant versus idigenous farmers. An analysis of factors affecting agricultural land use in the transitional agro-ecological zone of Ghana, 1984-2000. Danish Journal of geography, 106(1):103-113.

Dale, P.J. and Irwin, J.A. (1995) The release of transgenic plants from containment, and the move towards their widespread use in agriculture. Euphytica 85: 425-431, cited in Eastham, K. and Sweet, J. (2002).
Danso, A. A. and Morgan, P. (1993) Alley cropping maize (Zea mays var.Jeka) with cassia (Cassia siamea) in The Gambia: crop production and soil fertility, Agroforestry Systems, 21:2 (1993).
DeBeer, J. (2005). Reconciling Property Rights in Plants. J. World Intel. Prop. *8*, 5-31.
Debeljak, M., Ivanovska, A., Dzeroski, S., Meier-Bethke, S., Schiemann, J. (2007) Modelling spatial distribution of outcrossing rates between neighbouring maize fields In: Stein, A., J. and Rodriguez-Cerezo, E. (2007) Proceeding of the 3rd International Conference on Coexistence between Genetically Modified (GM) and non-GM based Agricultural Supply Chains (GMCC), Seville, Spain, 20-21 November 2007, Pp. 300-301.
deGrassi, A. (2003) Genetically Modified Crops and Sustainable Poverty Alleviation in Sub-Saharan Africa; An assessment of current evidence, Third World Network-Africa Publisher Website: www.twnafrica.org, accessed 18.6.2008)

Della Porta, G., Ederle, D., Bucchini, L., Prandi, M., Pozzi, C., Verderio, A. (2006) Gene Flow between Neighbouring Maize Fields in the Po Valley. A Fact-finding Investigation regarding Co-existence between Conventional and Non-Conventional Maize Farming in the Region of Lombardy, Italy. Publication of the Centro di Documentazione Agrobiotecnologie (CEDAB). http://www.cedab.it, accessed on 27.12.07.

DETR (1999). Investigation of feral oilseed rape populations. (authors: Charters YM, Robinson A, Squire GR). Genetically modified organisms research report No. 12. London UK: Defra.

Devaux C., Lavigne C., Austerlitz F., Klein E. K. (2007) Modeling and estimating pollen movement in oilseed rape (Brassica napus) at the landscape scale using genetic markers. Molecular Ecology 16: 487-499 (UPS).

D'Hertefeldt, T., Jørgensen, R. B., Pettersson, L. B. (2008) Long-term persistence of GM oilseed rape in the seedbank. Biol. Lett. 4, 314-317

Dietz-Pfeilstetter A., Metge K., Schönfeld J., Zwerger P. (2006) Assessment of transgene spread from oilseed rape by population dynamic and molecular analysis of feral oilseed rape. Journal of Plant Diseases and Protection, Special Issue XX: 39–47 (BBA).

Dima, S. J.; Ogunmokun, A. A ; Nantanga, T. (2002) The Status of Urban and Peri-urban Agriculture, Winkhoek and Oshakati, in Namibia. A survey report prepared for Integrated support to sustainable development and food security Programme (ip) in food and agriculture organization of the united Nations (FAO). ftp://ftp.fao.org/sd/sdw/sdww/nam_periurban_02.pdf, accessed on 7.05.07.

Directive 2001/18/EC of the European Parliament and of the Council of 12 March 2001 on the deliberate release into the environment of genetically modified organisms and repealing Council Directive 90/220/EEC.

Dubreuil, M., Warburton, M. Chastanet, D. Hoisington, A. Charcosset (2006) More on the introduction of temperate maize into Europe: Large-scale bulk SSR genotyping and new historical elements. Maydica 51: 281-291.

Eastham K. and Sweet, J. (2002) Genetically Modified Organisms (GMOs): The Significance of Gene Flow through Pollen Transfer. European Environment Agency (EEA) Environmental Issue Report No. 28. March 2002. Pp. 1-17. http://www.mindfully.org/GE/GE4/Pollen-Transfer-Gene-FlowEEAMar02.htm, accessed on 7.05.05.

Eder, J. (2006) Bericht zum Erprobungsanbau mit gentechnisch verändertem Mais in Bayern 2005, Schriftenreihe der Bayrischen Landesanstalt für Landwirtschaft, Freising Weihstephan, www.lfl.bayern.de

EFSA (European Food Safety Authority) EFSA GMO Risk Assessment FAQs, http://www.efsa.europa.eu/EFSA/efsa_locale-178620753816_EFSAGMORiskFAQs.htm, accessed on 12. 6.2008.

Egziabher, T. B.G. (2007) The Cartagena Protocol on Biosafety: History, Content and Implementation from a Developing Country Perspective *In:* Traavik, T. and Li Ching, L. (eds.) (2007) Biosafety first- Holistic approaches to risk and uncertainty in genetic engineering and genetically modified organisms. Tapit academic press. Pp. 389-405.

Ekboir, J., Boa, K. and Dankyi, A. A. (2002) Impact of No-Till Technologies in Ghana. Economics Program Paper 02-01. Mexico, D.F.: CIMMYT.

Emberlin J. (1999) A Report on the Dispersal of Maize Pollen (Summary). Environmental Impacts. http://www.biotech-info.net/maize_pollen.html, accessed on 13.04.07.

ETC Group (Action Group on Erosion, Technology and Conservation) (2003) Maize rage in Mexico. GM Maize Contamination in Mexico- 2 Years Later. http://www.etcgroup.org, accessed on 31.03.07.

European Commission [Online] Food Safety- From Farm to Fork http://ec.europa.eu/food/food/biotechnology/index_en.htm, accessed on 12.05.07.

European Commission GM Food and Feed- Community Register of GM Food and Feed. http://ec.europa.eu/food/dyna/gm_register/index_en.cfm.

European Union [Online] Food Safety/Health/Consumer Affairs. http://www.eurunion.org/policyareas/food.htm, accessed on 12.05.07.

Evenson, R. E. (2003) GMOs: Prospects for increased Crop Productivity in Developing Countries. Economic Growth Center. Yale University. Center Discussion Paper No. 878. http://www.econ.yale.edu/growth_pdf/cdp878.pdf, accessed on 23.04.2007.

Fabie, A. (2004) Research on coexistence in the field. French experiments for maize. COPA COGECA Colloquy on the co-existence and thresholds of adventitious presence of GMOs in conventional seeds, http://www.copa-cogeca.be/pdf/9.pdf

Fagan, J. (2007) Monitoring GMOs released into the environment and the food production system *In:* Traavik, T. and Li Ching, L. (eds.) (2007) Biosafety first- Holistic approaches to risk and uncertainty in genetic engineering and genetically modified organisms. Tapit academic press. Pp. 521-553.

Fakorede, M.A.B., Badu-Apraku, B., Kamara A.Y., Menkir, A., Ajala S.O. (2003) Maize revolution in West and Central Africa: An overview. International Institute for Tropical Agriculture (IITA).

FAO (2000) Maximum attainable crop yield ranges for high and intermediate level inputs in tropical, sub-tropical and temperate environments, http://www.fao.org/ag/agl/agll/gaez/tab/t38.htm, accessed on 24.02.07.

FAO (2007a) Zea mays L. http://www.fao.org/ag/agp/agpc/doc/gbase/data/pf000342.htm, accessed on 7.05.07.

FAO (2007b) FAOSTAT. http://faostat.fao.org/site/336/DesktopDefault.aspx?PageID=336, accesed on 24.02.2007.

FAO (2008) Genetic Resources http://www.fao.org/biodiversity/geneticresources/en/, accessed on 18.6.2008.

FoEE (Friends of the Earth Europe) FoEE Biotechnology Programme and European GMO Campaign [Online] http://www.foeeurope.org/GMOs/european_legislation/genetically_modified.htm, accessed on 12.05.07.

FoEE (Friends of the Earth) (2008) GMO contamination in Nicaragua, http://www.foe.org/foodaid/, accessed 19.6.2008.

Fosu, M., Kühne, R. F. and Vlek P. L. G. (2004) Improving Maize Yield in the Guinea Savannah Zone of Ghana with Leguminous Cover Crops and PK Fertilization. Journal of Agronomy 3 (2): 115-121.

GAIN (Global Agriculture Information Network) (2005) Ghana Biotechnology Agricultural Biotechnology. USDA Foreign Agricultural Service Report. Pp. 3-9.

Galinat, W.C. (1988). The origin of corn. In: Corn and Corn Improvement. Agronomy Monographs No.18. American Society of Agronomy, G.F Sprague and J.W. Dudley, (eds.). Madison, Wisconsin, pp. 1-31.

Garmin, ETREX GPS Reciever, http://www.garmin.com/garmin/cms/site/us, accessed 10.06.2008

Garnier A., David O., Deville A., Lecomte J., Laredo C. (In press). Estimation of demographic parameters for populations with immigration and incomplete observations, feral oilseed rape as a case study (UPS, INRA).

Gaugitsch, H. (2004) Environmental Risk Safety Assessment of GMOs- Methods and Criteria In: Breckling, B. & Verhoeven (eds): Risk Hazard Damage- Specification of Criteria to Assess Environmental Impact of Genetically Modified Organisms. Bonn (Bundesamt für Naturschutz). Naturschutz und Biologische Vielfalt 1. Pp.185-193.

GENET (2005) Ghana stops importation of GM foods, published 28 Jul 2005 http://www.gene.ch/genet/2005/Aug/msg00004.html, accessed 30.6.2008.

Ghana districts. Demographic Characteristics in the Greater Accra Region [Online] http://www.ghanadistricts.com/districts/?r=1&_=1&sa=3823, accessed on 20.04.08.

Ghana National Population and Housing Census (2000), Accra.

Ghanaweb (2005) Workshop on Genetically Modified Organisms. http://www.ghanaweb.com/GhanaHomePage/NewsArchive/artikel.php?ID=86822, accessed on 11.10.05.

GMWatch (2008) USA - StarLink maize - a GM maize intended for animal feed was found in human food, http://gmcontaminationregister.org/index.php?content=re_detail&gw_id=11®=reg.2&inc=1&con=0&cof=1&year=0&handle2_page=, accessed 30.6.2008.

Google maps, http://maps.google.com, accessed 01.5,2008.

Gray, A. J. & Raybould, A. F. (1998) Reducing transgene escape routes, Nature 392, 653–654.

Greenpeace (2003) Genetically Engineered (GE) Papaya-Unknown Plant. June 2003. http://www.greenpeace.org/international_en/multimedia/download/1/290394/0/papaya_unknown_plant.pdf, accessed on 4.05.07.

Greenpeace (2006) Illegal GM rice found in the UK, http://www.greenpeace.org.uk/tags/gm-contamination, accessed 19.6.2008.

Greenpeace (2007) Greenpeace moves to block GMO rice approval in the Philippines http://www.greenpeace.org/seasia/en/press/releases/greenpeace-moves-to-block-gmo, accessed 18.6.2008

Greenpeace (2008) Argentina - illegal planting of Monsanto GM maize discovered, Contamination Register, http://www.gmcontaminationregister.org/index.php?content=re_detail&gw_id=44®=cou.22&inc=0&con=0&cof=1&year=0&handle2_page=, accessed on 19.6.2008.

Greenpeace (2008) Germany – cross-pollination by GM maize of neighbouring crop GM Contamination Register, http://gmcontaminationregister.org/index.php?content=re_detail&gw_id=100®=cou.5&inc=0&con=0&cof=1&year=0&handle2_page=, accessed 19.6.2006.

Greenpeace (2008) Greece – maize seed contamination, Contamination Register, http://www.gmcontaminationregister.org/index.php?content=re_detail&gw_id=146®=cou.26&inc=0&con=0&cof=1&year=0&handle2_page=, accessed on the 19.6.2008

Greenpeace (2008) Japan – Bt10 maize detected in imports, GM Contamination Register, http://gmcontaminationregister.org/index.php?content=re_detail&gw_id=81®=cou.10&inc=0&con=0&cof=1&year=0&handle2_page=, accessed 19.6.2008.

Greenpeace (2008) Kenya – Unapproved GM maize (MON 810) found in seed imports, GM Contamination Register, http://gmcontaminationregister.org/index.php?content=re_detail&gw_id=234®=cou.57&inc=0&con=0&cof=1&year=0&handle2_page=, accessed 19.6.2008.

Greenpeace (2008) Philippines - Farmers lured into planting Bt maize, GM Contamination Register, http://gmcontaminationregister.org/index.php?content=re_detail&gw_id=130®=cou.38&inc=0&con=0&cof=1&year=0&handle2_page=, accessed 19.6.2008.

Grosse, N. Beckmann, V., Schleyer, C. (2008) The role of coordination and corporation for growing of GM crops: The case of Bt-maize in Brandenburg, Germany. Poster at GMLS 2008, Bremen (www.gmls.eu), proceedings in preparation.

Gupta, P.K. and Varshney, R. K. (2000) The Development and Use of Microsatellite markers for genetic analysis and plant breeding with emphasis on bread wheat. Euphytica 113: 163-185.

Haygood, R.; Ives, A. R.; Andow, D. A. (2003) Consequences of Recurrent Gene Flow from Crops to Wild Relatives. Proceedings of the Royal Society B: Biological Sciences. 270: 1527. Pp. 1879-1886.

Heinemann, J. A. (2007) A typology of the effects of trans(gene) flow on the conservation and sustainable use of genetic resources. Rome UN FAO: 1-94.

Henry, C., Morgan, D., Weekes, R., Daniels, R., & Boffey, C. (2003) Farm scale evaluations of GM crops: monitoring gene flow from Gm crops to non-GM equivalent crops in the vicinity (contract reference EPG 1/5/138). Part I: Forage Maize. Final report, 2000/2003.

Herrera, J. C., Combes, M. C., Cortina, H., Alvarado, G., Lashermes, P. (2002) Gene Introgression into Coffea arabica by way of Triploid Hybrids (C. Arabica C. canephora). Heredity (2002) 89, 488-494.

Hewlett, K. and Azeez, G. (2007) The Economic Impacts of GM Contamination Incidents on the Organic Sector In: Proceedings of the Third International Conference on Coexistence between Genetically Modified (GM) and non-GM based Agricultural Supply Chains, Seville (Spain), 20/21 November 2007. pp. 336-337.

Hilbeck, A., Nelson, K., Andow, D. A., Underwood, E. (2004) Am Scientist's use of Problem Formulation and Options Assessment (PFOA) in Risk Assessment of GM Crops In: Breckling, B. & Verhoeven (eds): Risk Hazard Damage- Specification of Criteria to Assess Environmental Impact of Genetically Modified Organisms. Bonn (Bundesamt für Naturschutz). Naturschutz und Biologische Vielfalt 1. Pp. 131-143.

Hitchcock, A. S. and Chase A. (1951) manual of the Grasses of the United States (Volume 2). Dover Publications: N.Y. p. 790-796.

Hofmann, F. (2007) Kurzgutachten Zur Abschätzung Der Maispollendeposition In Relation Zur Entfernung Von Maispollenquellen Mittels Technischem Pollensammler Pmf, Gutachten „Maispollendeposition“, Bundesamt für Naturschutz BfN, http://www.oekologiebuero.de/Gutachten-BfN-Maispollendeposition.pdf, accessed 22.6.2008.

Holck, A., Vaitilngom, M., Didierjean, L., Rudi, K. (2002) 5 '-Nuclease PCR for Quantitative Event-specific Detection of the Genetically Modified MON810 MaisGard Maize. European Food Research and Technology. 214(5):449-453.

Hovorka, A. J. and Lee-Smith, D. (2006) Gendering the Urban Agriculture Agenda In: van Veenhuizen, R. (2006) Cities Farming for the Future. Urban Agriculture for Green and Productive Cities. http://www.ruaf.org/node/961, accessed on 12.05.07.

IDRC (The International Development Research Center) Technology Adoption by Small-Scale Farmers in Ghana, http://www.idrc.ca/en/ev-30794-201-1-DO_TOPIC.html, accessed 2.7. 2008.

IITA (International Institute for Tropical Agriculture) (2007). http://www.iita.org, accessed on 06.01.2008.

ISAAA (International Service for the Acquisition of Agri-Biotech Applications) (2006) Global Status of Commercialized Biotech/GM Crops. http://www.isaaa.org/resources/publications/briefs/35/pptslides/default.html, accessed on 31.3.07.

IUCN (2000) IUCN-CEESP Environment and Security Task Force Briefing, Presented at the World Conservation Conference, Amman.

Jemison, J.M., & Jr.Vayda, M.E. (2001) Cross pollination from genetically engineered corn: wind transport and seed source. AgBioForum. 4: 87-92.

Johannessen M. M., Andersen B. A., Jørgensen R.B. (2006) Competition affects the production of first backcross offspring on F1 hybrids Brassica napus x B. rapa. Euphytica 150: 17-25 (Risø).

Jones, C.J., Edwards, K.J., Castaglione, S., Winfield, M. O., Sala, F., van de Wiel, Bredemeijer, G., Vosman, B, Matthes, M., Daly, A., Brettschneider, R., Bettini, P., Buiatti, M., Maestri, E., Malcevschi, A., Marmiroli, N., Aert, R., Volckaert, G., Rueda, J., Linacero, R., Vazquez, A., Karp, A. (1997) Reproducibility testing RAPD, AFLP and SSR markers in plants by a network of European laboratories. Molecular Breeding 3: 381-390.

Jones, M. D. and Brooks, J. S. (1952) Effect of tree barriers on outcrossing in corn, Oklahoma Agricultural Experimental Station, Technical Bulletin No. T-45.

Jones, M.D. and Brooks, J. S. (1950) Effectiveness of distance and border rows in preventing outcrossing in corn. Oklahoma Agricultural Experimental Station, Technical Bulletin No. T-38.

Jurgenheimer R. (1976) in Corn improvement, seed production and uses, Wiley Interscience In Henry C et al (2003) Farm scale evaluations of GM crops: monitoring gene flow from GM crops to non GM equivalents in the vicinity: part one forage maize, DEFRA report. EPG/1/5/138.

Katul, G. G., Porporato, A., Nathan, R., Siqqueira, M., Soons, M. B., Poggi, D., Horn, H. S., Levin, S. A. (2005) Mechanistic analytical models for long-distance seed dispersal by wind. The American Naturalist, 166:368-381.

Kawata (2008) Dispersal and persistence of genetically modified oilseed rape around Japanese harbours, Proceeding of the International Conference on Implications of GMO cultivation at Large Spatial Scales, 2-4th April, 2008, Bremen.

Klein, E. K., Laredo, C., Lavigne, C. (1998) Estimation of pollen dispersal function from field experiments, cited in Brookes et al. (2004).

Knispel, L. K, McLachlan, S. M., Van Acker, R. C., Friesen, L. F. (2007) Multiple herbicide tolerance in escaped canola populations in Manitoba, Canada. Proceedings of the International Conference on the Coexistence of GM and Non-GM Crops, Seville, Spain, Pp. 61-64.

Kuparinen, A. (2006) Gene Flow from Transgenic Plant Populations: Models and Applications for Risk Assessment. PhD Dissertation, Department of Mathematics and Statistics, Faculty of Science, University of Helsinki.

Kyriakidou D. (2000) Greece to Further Test, Destroy and GM Cotton Crops. Reuters News Service. July 4, 2000.
http://www.planetark.org/dailynewsstory.cfm?newsid=7343, accessed on 4.05.07.

Laue, H., O. A. (2004) Automated Detection of Canola/Rapeseed Cultivation from Space. Application of new Algorithms for the identification of Agriculrural Plants with Multispectral Satellite Data on the example of Canola Cultivation, PhD thesis at the Institute of Environmental Physics, University of Bremen.

Lecomte, J., Jørgensen R., Bartkowiak/Broda, I., Devaux, C., Dietz-Pfeilstetter, A., Gruber, S., Husken, A, Kuhlmann, M., Lutman, P., Rakousky, S., Sausse, C., Squire, G., Sweet, J., and Aheto, D.W. (2007) Gene Flow in Oilseed rape: What the datasets of the SIGMEA EU Project tell us for coexistence. Proceedings of the International Conference on the Coexistence of GM and Non-GM Crops, Seville, Spain. Pp. 49-52.

Lecomte, J., Jørgensen R., Bartkowiak/Broda, I., Devaux, C., Dietz-Pfeilstetter, A., Gruber, S., Husken, A, Kuhlmann, M., Lutman, P., Rakousky, S., Sausse, C., Squire, G., Sweet, J., and Aheto, D.W. (2007) Gene Flow in Oilseed rape: What the datasets of the SIGMEA EU Project tell us for coexistence. Proceedings of the International Conference on the Coexistence of GM and Non-GM Crops, Seville, Spain. Pp. 49-52.

Lecoq, E., Holt, K., Janssens, G., Legris, A., Pleysier, B., Tinland and Wandelt, C. (2007) General Surveillance: Roles and Responsibilities. The Industry View, J. Verbr. Lebensm. 2 (2007) Supplement 1:25-28.
Lohr, H (1999) Sampling: Design and Analysis, Duxbury, ISBN 0-534-35361-4
Lombardy, Italy. Report, Centro Documentazione Agrobiotechnologie, Milan, Italy.
Losey, J. E., Rayor, L.S., Carter, M.E. (1999) Transgenic Pollen Harms Monarch Larvae. Nature 399, 214.
Loubet, B and Foueillassar, X. (2003) INRA Thiverval-Grignon Etude mécaniste du transport et du dépôt de pollen de maïs dans un paysage hétérogène.Rapport de fin de projet Convention INSU N° 01 CV 081.
Lutman P., Berry K., Payne R., Sweet J., Simpson E., Law J., Walker K., Wightman P., Champion G., May M., Lainsbury M. (2005) Persistence of seeds from crops of conventional and genetically modified herbicide tolerant oilseed rape (Brassica napus) Proceedings of the Royal Society B, 272, 1909-1915 (RRes, NIAB).
Lyon, F. and Afikorah-Danquah, S. (1998) Small-scale Seed Provision in Ghana: Social Relations, Contracts and Institutions for Micro-Enterprise Development, ODI Agricultural Research and Extension Network (AgREN). Network Paper, No. 84. July 1998.

Ma, B.L., Subedi, K.D. & Reid, L.M. (2004) Crop ecology, management & quality. Extent of
Magorokosho, C. (2006) Genetic Diversity and Performance of Maize Varieties from Zimbabwe, Zambia and Malawi, A PhD Dissertation, Texas A and M University. December, 2006.
Mangelsdorf, P. C. (1974). Corn: its Origin, Evolution and Improvement. Harvard Univ. Press, Cambridge, Mass.
Map of West Africa, http://www.cpj.org/Briefings/2000/Bekoutou/map.html, accessed on 15.6.2007.
Marsan, P. A., Castiglioni, P., Fusari, F., Kuiper, M., and Motto, M. (1998) Genetic Diversity and its Relationship to Hybrid Performance in Maize as Revealed by RFLP and AFLP markers. Theor Appl Genet, 96: 219- 227.
Maxwell, D. and Armar-Klemesu, M. [Online] Urban Agriculture in Greater Accra: ReviewingResearch Impacts for Livelihoods, Food, and Nutrition Security. International Development Research Centre (IDRC). Canada. http://www.idrc.ca/fr/ev-8303-201-1-DO_TOPIC.html#IVINTRO, accessed on 12.05.07.
Meir-Bethke & Schiemann, J. (2003) Effect of varying distances and intervening maize fields on outcrossing rates of transgenic maize, Proceedings of the 1st European conference on the co-existence of GM crops with conventional and organic crops, Denmark, November 2003.
Melé, E. (2004) Spanish study shows that coexistence is possible. Agricultural Biotechnology International Conference ABIC 3, 2.

Menzel G. (2006) Verbreitungsdynamik und Auskreuzungspotenzial von Brassica napus L. (Raps) im Großraum Bremen. Basiserhebung zum Monitoring von Umweltwirkungen transgener Kulturpflanzen (Dispersal dynamics and outcrossing potential of Brassica napus L. (Oilseed rape) in the Region of Bremen (Baseline Survey for the Monitoring of the Environmental Impact of Cultivated Transgenic Plants). PhD thesis, Universität Bremen. Forschen und Wissen- Umweltwissenschaft. GCA-Verlag, Waabs. PhD Dissertation, University of Bremen.

Messĕan, A. (1999) Impact du development des plantes transgeniques dans les systemes de culture: rapport final. Dossier No. 96/15- B: Impact des plantes transgeniques. http://www.acta.asso.fr/cr/cr9924.htm.

Messĕan, A. and Angevin, F. (2007) Coexistence measures for maize cultivation: lessons from gene flow and modelling studies. Proceedings of the International Conference on the Coexistence of GM and Non-GM Crops, Seville, Spain. Pp. 23-26.

Messeguer, J. (2004) Gene Flow Assessment in Transgenic Plants. Biomedical and Life Sciences. Plant Cell, Tissue and Organ Culture. 73:3. Pp. 201-212.

Messeguer, J. Penås, G., Ballester, J., Bas, M., Serra, J., Salvia, J., Palausdelmås, M. & Melė, E. (2006) Plant Biotechnology Journal, 4:633-645.

Meyer, H. (2005) German Policies on Biosafety in regard to the African Union, March 2005, Berlin. Pp. 1-5.

Morris, L. M., Tripp, R., and Dankyi, A. A. (1999) Adoption and Impacts of Improved Maize Production Technology: A Case Study of the Ghana Development Project. Economics Program Paper 99-01. Mexico, D.F.: CIMMYT.

Morris, L. M., Tripp, R., and Dankyi, A. A. (1999) Adoption and Impacts of Improved Maize Production Technology: A Case Study of the Ghana Development Project. Economics Program Paper 99-01. Mexico, D.F.: CIMMYT.

Muhammed, L. and Underwood, E. (2004) The Maize Agricultural Context in Kenya In: Hilbeck, A. and Andow, D.A. (eds) (2004) Environmental Risk Assessment of Genetically Modified Organisms Volume 1: A Case Study of Bt Maize in Kenya. CABI Publishing, Wallingford, UK.

Myers, N. (1997) Environmental refugees. Population and Environment 19:167-182.

National Biosafety Framework for Ghana [editors Owusu-Biney, et al.]. - Geneva: United Nations Environment Program, Global Environment Facility ; Accra : The Ministry of Environment and Science: Biotechnology & Nuclear Agriculture, Research Institute, 2004.
http://www.unep.ch/biosafety/development/countryreports/GHNBFrep.pdf, accessed on 11.10.06.

Network of Concerned Farmers (2004) NSW Minister guarantees farmers are protected, http://www.non-gm-farmers.com/contact.asp, accessed 29.6.2008.

Nuffield Council on Bioethics (1999) Genetically Modified Crops: Social and Ethical Issues. London.

Obosu-Mensah K. 'Changes in Official Attitudes Towards Urban Agriculture in Accra'. African Studies Quarterly 6, No. 3. http://web.africa.ufl.edu/asq/v6/v6i3a2.htm., accessed on 7.05.07.

Ortega Molina, J. 2004. Results of the studies into the coexistence of genetically modified and conventional maize. COPA-COGECA colloquy on the co-existence and thresholds of adventious presence on GMOs in conventional seeds, http://www.copa-cogeca.be/pdf/9.pdf

Palaudelmås, M., Messeguer, J., Penas, G., Serra, J., Salvia, J., Plan, M., Nadal, A., and Mele, E. (2007) Effect of sowing and flowering dates on maize gene flow In: Stein, A., J. and Rodriguez-Cerezo, E. (2007) Proceeding of the 3rd International Conference on Coexistence between Genetically Modified (GM) and non-GM based Agricultural Supply Chains (GMCC), Seville, Spain, 20-21 November 2007. Pp. 235-236.

Papa, R. (2005) Gene Flow and Introgression between Domesticated Crops and their Wild Relatives. The Role of Biotechnology. Villa Gualino, Turin, Italy. 5-7 March 2005. http://www.fao.org/biotech/docs/papa.pdf, accessed on 7.05.07.

Pengue W. A. (2004) Environmental and Socio-economic impacts of transgenic crops in Argentina and South America: An ecological economics approach In: Breckling, B. & Verhoeven (eds): Risk Hazard Damage- Specification of Criteria to Assess Environmental Impact of Genetically Modified Organisms. Bonn (Bundesamt für Naturschutz) Naturschutz und Biologische Vielfalt 1. Pp. 49-59.

Pessel, F. D., Lecomte, J., Emeriau, V., Krouti, M., Messean, A. & Gouyon, P.H. (2001) Persistence of Oilseed rape (Brassica napus L.) outside of cultivation fields. Theor. Appl. Genet. 102, S. 841-846.

Pivard, S. (2006) Processus écologiques impliqués dans la présence d'une plante cultivée dans les milieux naturels et semi-naturels d'un agroécosystème : Cas des populations férales de colza (Brassica napus L.). Thèse de l'Univ. Paris-Sud, 194 pp.

Pivard, S., Adamczyk K, Lecomte J, Lavigne C, Bouvier A, Deville A, Gouyon PH, Huet S. (in press) Where do the feral oilseed rape populations come from? A large-scale study of their possible origin in a farmland area. Journal of Applied Ecology. (UPS, INRA).

Planetdiversity (2008) Together for a diverse future- World Congress on the future of Food and Agriculture, Bonn, 12-18 May, 2008. http://www.planet-diversity.org/why.html, accessed 18.6.2008.

Pleasants, J. M., Hellmich, R. L., and Lewis, L. C. (1999) Pollen Deposition in Milkweed Leaves under Natural Conditions (Presentation at the Monarch Butterfly Research Symposium, Chicago.

Purseglove, J. W. (1972) Tropical Crops: Monocotyledons 1. Longman Group Limited., London.

Quist, D. (2004) Transgene Ecology: An Ecological Perspective for GMO Risk Assessment In: Breckling, B. & Verhoeven (eds): Risk Hazard Damage- Specification of Criteria to Assess Environmental Impact of Genetically Modified Organisms. Bonn (Bundesamt für Naturschutz), Naturschutz und Biologische Vielfalt 1. Pp. 239-243.

Quist, D. and Chapela, I.H. (2001) Transgenic DNA introgressed into traditional maize land races in Oaxaca, Mexico. Nature 414:541-543.

Radu, A, Urechean, V., Naidin, C. & Motorga, V. (1997). The theoretical significance of a mutant in the phylogeny of the species Zea mays L. Maize Newsletter 71, 77-78.

Reuter, H.; Breckling, B.; and Böckmann, S. (2008) Modelling maize hybridisation on a landscape level. Data analysis and development of a dispersal kernel. Proceedings, GMLS 2008, Bremen (www.gmls.eu).

Rosendal, G. K., Olesen, I., Bentsen, H. B., Tvedt, M. W. and Bryde, M. (2006). Access to and Legal Protection of Aquaculture Genetic Resources—Norwegian Perspectives. J. World Intell. Prop. *9*, 392-412.

RUAF (Resource Centres on Urban Agriculture and Food Security) Urban Agriculture in Accra (Ghana) [Online]. http://www.ruaf.org/node/498, accessed on 7.05.07.

Sabellek, K.; Lipsius, K.; Richter, O. and Wilhelm, R. (2007) Influence of flowering heterogeneity on cross-pollination rates in maize: experiments and modeling In: Stein, A., J. and Rodriguez-Cerezo, E. (2007) Proceeding of the 3rd International Conference on Coexistence between Genetically Modified (GM) and non-GM based Agricultural Supply Chains (GMCC), Seville, Spain, 20-21 November 2007. Pp. 265-266.

Salamov, A. B. (1940) About isolation in corn. Sel. I. Sem., 3. Russian translation by Michael Afanasiev in 1949.

Schiemann, J. (2003) Co-existence of genetically modified crops with conventional and organic farming. Environmental Biosafety Research, 2:213-217.

Schmeiser P. (1999) Monsanto vrs. Schmeiser. The Classic David and Goliath Struggle. http://www.percyschmeiser.com/, accessed on 4.05.07.

Sears, M. K. and Stanley-Horn, D. (2000) Impact of Bt Corn Pollen on Monarch Butterfly Populations In: Fairbairn, C., Scoles, G. and McHughen, A. (eds) Proceedings of the 6th International Symposium on the Biosafety of Genetically Modified Organisms. University Extension Press, Canada.

See GMOs, Biofuels and the Third World [Online]. Independent Media Centre. Portland. http://portland.indymedia.org/en/2005/12/329780.shtml, accessed on 12.05.07.

SeedQuest (2005) Ghana Strongly Favours GM Crops. SciDev.Net, News Section. http://www.seedquest.com/News/releases/2005/august/13219.htm, accessed on 23.04.07.

Senior, M. L. and Heun, M. (1993) Mapping maize microsatellites and polymerase chain reaction confirmation of the targeted repeats using a CT primer. Genome, 36: 884-889.

Seralini (2005) MON 863: GM Maize Produced by Monsanto Company. Greenpeace Report. Pp.1-5.

SIGMEA (2008) Field/feral/volunteer/wild relative demography. WP2 Final Project Report (In the thematic priority Policy-oriented research).

Simpson, E. C., Norris, C. E., Law, J. R., Thomas, J. E. & Sweet, J. B. (1999) Gene flow in genetically modified herbicide tolerant oilseed rape (Brassica napus) in the UK. In: Gene Flow and Agriculture: Relevance for Transgenic Crops. Lutman, P. (Ed.). BCPC Symposium Proceedings No. 72.

Smit, J. and Bailkey, M. (2006) Urban Agriculture and the Building of Communities. RUAF, http://www.ruaf.org/node/971, accessed on 7.05.07.

Snow A. A., Andow DA, Gepts P, Hallerman EM, Power A, Tiedje JM, and Wolfenbarger LL (2005). Genetically engineered organisms and the environment: Current status and recommendations. Ecological Applications, 15: 377–404.

Soleri, D., Cleveland, D. A., Aragon, F. C., Fuentes, M. R. L, Rios, H. L., Sweeney, S. H. (2005) Understanding the Potential Impact of Transgenic Crops in Traditional Agriculture: Maize Farmers' Perspectives in Cuba, Guatemala and Mexico. Environ. Biosafety Res. 4 (2005). Pp. 141-166.

Taramino, G. and Tingey, S. (1996) Simple Sequence Repeats for Germplasm Analysis and mapping in maize. Genome, 39: 277-287.

Thompson, C. E., Squire, G., Mackay, G. R., Bradshaw, J. E., Crawford, J., Ramsay, G. (1999) Regional patterns of gene flow and its consequences for GM oilseed rape. In: Lutman, PJW (Chair) Gene flow and Agriculture. Relevance for Transgenic Crops,. Symp Proc. No. 72. British Crop Protection Council, Farnham, Surrey: pp. 95-100.

Traavik, T. (2008) GMOs and their Unmodified Counterparts: Substantially Equivalent or Different? GMLS Conference Proceedings Bremen, 2-4th April, 2008.

Treu, R. and Emberlin, J. (2000) Pollen Dispersal in the Crops Maize (Zea mays), oilseed rape (Brassica napus ssp oleifera), Potato (Solanum tuberosum), Sugar beet (Beta vulgaris ssp vulgaris) and Wheat (Triticum aestivum). Report for the Soil Association. http://www.soilassociation.org/web/sa/saweb.nsf/b0062cf005bc02c180256a6b003d987f/80256ad80055454980256866075e5b4!OpenDocument&Highlight=2,Treu, accessed on 13.05.07.

Tripp, R. and Marfo, K. (1997) Maize Technology Development in Ghana during Economic Decline and Recovery In: Byerlee D., Eicher C. K. (eds) Africa's Emerging Maize Revolution. Boulder, Colorado: Lynne Rienner Publishers.

Tsimese, L. K. (2003) Dumping GMOs in Africa. Agricultural Reform Movement (Ghana), http://www.greens.org/s-r/32/32-05.html, accessed 18.6.2008

UNDP (1996). Urban Agriculture. Food, Jobs and Sustainable Cities. United Nations Development Programme. Publication Series for Habitat II. Vol. 1; New York, USA.

Van Veenhuizen, R. (2006) Cities Farming for the Future- Urban Agriculture for Green and Productive Cities. RUAF (2006). http://www.ruaf.org/node/961, accessed on 7.05.07.

Walker, D.J., Tripp, R., Opoku-Apau, A. (1998) Seed Management by Small-scale Farmers in Ghana. A Study of Maize and Cowpea Seed in the Brong-Ahafo and Volta Regions. Natural resources Institute.

Walsh, P. S., Metzger, D. A. And Higuchi, R. (1991) Chelex® 100 as a Medium for Simple Extraction of DNA for PCR-Based Typing from Forensic Material. Research Report, Biotechniques, Vol. 10, No.4.

Walsh, P.S., Fildes, N., Louie, A., Higuchi, R. (1991) Report of the blind trial of the Cetus AmpliTypeTM HLA DQα forensic DNA amplification and typing kit. J Forensic Sci.

Warburton, M. L., Xianchun X., Crossa J., Franco J., Melchinger, A. E; Frisch M.; Bohn M., and Hoisington D. (2002) Genetic Characterization of CIMMYT Inbred Maize Lines and Open Pollinated Populations Using Large Scale Fingerprinting Methods Crop Sci. 42:1832–1840.

Watson, L. and Dallwitz, M. J. (1992) Grass Genera of the World: Descriptions, Illustrations, Identification, and Information Retrieval; including Synonyms, Morphology, Anatomy, Physiology, Phytochemistry, Cytology, Classification, Pathogens, World and Local Distribution, and References. Version: 18th August 1999. http://biodiversity.uno.edu/delta.

Weber, W. E., Bringezu, T., Broer, I., Holz, F. & Eder, J. (2005) Koexistenz von gentechnisch verändertem und konventionellem Mais. mais – Die Fachzeitschrift für den Maisanbauer. Sonderdruck 1+2/2005.

Weber, W.E.; Bringezu, T.; Broer, I.; Eder, J.; Holz, F. (2007) Coexistence between GM and non-GM maize crops – tested in 2004 at the field scale level. J. Agronomy & Crop Sciences. 193: 79-92.

Wikipedia (2007) Urban Agriculture. http://en.wikipedia.org/wiki/urban_agriculture, accessed on 17.05.07.

Windels, P., Taverniers, I., Depicker, A., Van Bockstaele, E., De Loose, M., (2001) Charactirization of the Roundup Ready Soybean Insert. European Food Research and Technology. 213 (2):107-112.

World Bank (2005) Indigenous innovation in farmer-to-farmer extension in Burkina Faso. Indigenous Knowledge Notes, http://www.worldbank.org/afr/ik/iknt77.htm, accessed 18.6.2008.

Ye-yun, X, Zhan, Z., Yi-ping, X, Long-ping, Y. (2005) Identification and Purity Test of Super Rice with SSR Molecular Markers. Rice Science 2005, 12 (1): 7-12.

ZEF, Zentrum für Entwicklungsforschung (2002/2003) Research on Genetically Modified Cotton in India In: Annual Report 2002/2003. Bonn.

Appendices

Appendix 1: Study background

Appendix 1.1: Transgenic food and feed products of maize as submitted for notification to the European Commission of 2007.

Transformation event (Unique ID)	Specification of the project
MON 863 (MON-ØØ863-5)	Foods and food ingredients derived from genetically modified maize (Zea maize L.) line MON 863 with increased protection to insects and from all its crosses with traditionally bred maize lines. MON 863 maize contains a modified cry3Bb1 gene derived from Bacillus thuringiensis subsp. kumamotoensis, which confers resistance to the corn rootworm Diabrotica spp., under the regulation of a promoter derived from Cauliflower Mosaic Virus, with translation enhancer from wheat (Triticum aestivum).
MON 810 (MON-ØØ81Ø-6)	The insect-protected maize line was generated by particle acceleration technology using plasmids. The transgenic maize line produced expresses the cry1A(b) gene (origin - Bacillus thuringiensis subsp. kurstaki) which encodes a cry1A(b) insect control protein specific to lepidopteran pests.
MON 863 x MON 810 (MON-ØØ863-5 x MON-ØØ81Ø-6)	The maize hybrid MON 863 x MON 810 was produced by conventional breeding to combine the rootworm resistance trait in MON 863 with the lepidopteran insect resistance trait present in GM maize, MON 810.
NK 603 (MON-ØØ6Ø3-6)	Foods and food ingredients derived from genetically modified maize (Zea maize L.) line NK603 with increased tolerance to the herbicide glyphosate and from all its crosses with traditionally bred maize lines.
MON863 x MON603 (MON-ØØ863-5 x MON-ØØ6Ø3-6)	Hybrid maize MON863 x NK603 produced by conventional breeding to combine the rootworm resistance trait in MON 863 and glyphosate resistance trait in NK603
NK 603 x MON 810 (MON-ØØ6Ø3-6 x MON-ØØ81Ø-6)	Hybrid maize NK603 X MON810 is produced by a single traditional cross of NK603 maize and MON810 maize inbred lines (homozygous for the respective introduced trait). The intended functions of the genetic modifications are tolerance to glyphosate (NK603) and resistance to lepidopteran larvae of Sesamia spp. and Ostrinia nubialis (MON810).
DAS1507 (DAS-Ø15Ø7-1)	Foods and food ingredients containing, consisting of, or produced from genetically modified maize (Zea maize L.) line 1507, with resistance to the European corn borer (Ostrinia nubilalis) and certain other lepidopteran pests and with tolerance to the herbicide glufosinate-ammonium.
GA21 (MON-ØØØ21-9)	Foods and food ingredients produced from genetically modified maize (Zea mays L.) line GA21 with increased tolerance to the herbicide glyphosate and from all its crosses with traditionally bred maize lines.
T25 (ACS-ZMØØ3-2)	Genetically modified maize (Zea mays L.) with increased glufosinate ammonium tolerance derived from the maize line HE/89 transformation event T25 which has been transformed using plasmids.

Transformation event (Unique ID)	Specification of the project
Bt11 (SYN-BT Ø11-1)	Sweet maize, fresh or canned, that is progeny from traditional crosses of traditionally bred maize with genetically modified maize line Bt11 that contains a synthetic version of the cryIA (b) gene derived from a Bacillus thuringiensis kurstaki strain.
Bt176 (SYN-EV176-9)	Genetically modified maize (Zea mays L.) with the combined modification for insecticide properties conferred by the Bt-endotoxin gene and increased tolerance to the herbicide glufosinate ammonium. The product consists of inbred lines and hybrids derived from a maize (Zea mays L.) line (CG 00256-176).
GA21 x Mon 810 (MON-ØØØ21-9 x MON-ØØ81Ø-6)	GA21 x MON 810 maize is the hybrid of maize line GA 21 and maize line MON 810.

Source: European Commission, GM Food and Feed- Community Register of GM Food and Feed (2007).

Appendix 1.2: Maize varieties and hybrids developed under the Ghana Grains Development Project (GGDP).

Name of variety	Year of release	Grain colour	Grain texture	Maturity (days of flowering)	Yield (t/ha)	Streak resist-ant?	Nutrition-ally en-hanced?	CIMMYT germplasm
Aburotia	1984	White	Dent	105	4.6	No	No	Tuxpeno PBC16
Dobidi	1984	White	Dent	120	5.5	No	No	Ejura (1) 7843
Kawanzie	1984	Yellow	Flint	95	3.6	No	No	Tocumen (1) 7931
Golden crystal	1984	Yellow	Dent	110	4.6	No	No	-
Safita-2	1984	White	Dent	95	3.8	No	No	Pool 16
Okomasa	1988	White	Dent	120	5.5	Yes	No	EV8343-SRa
Abeleehi	1990	White	Dent	105	4.6	Yes	No	Ikenne 8149-SRa
Dorke SR	1990	White	Dent	95	3.8	Yes	No	Pool 16-SRa
Obatanpa	1992	White	Dent	105	4.6	Yes	Yes	Pop 63-SRa
Mamabab	1996	White	Flint	110	6.0	Yes	Yes	Pop. 62, Pop. 63-SRa
Dadabab	1996	White	Dent/ Flint	110	6.0	Yes	Yes	Pop. 62, Pop. 63-SRa
Cidabab	1996	White	Dent	110	6.0	Yes	Yes	Pop. 62, Pop. 63-SRa

Source: Morris et al, 1999 ('a' means developed jointly with IITA. SR= resistant to maize streak virus; 'b' means Three-way cross hybrid).

Appendix 1.3: Parameter descriptors for commercial cultivars in Accra.

Code	Cultivar	Seed type
21	Obatanpa	Improved variety
22	Mamaba	Improved variety
23	Abeleehi	Improved variety
24	Probable Obatanpa/ Mamaba	Improved variety
25	Local variety (Traditional)	Local hybrid
26	Okomasa	Improved variety
27	Probable Obatanpa & local	Mixed varieties (Improved & Local)
28	Aburotia	Improved variety
29	Dobidi	Improved variety
30	Kawanzie	Improved variety
31	Golden crystal	Improved variety
32	Safita-2	Improved variety
33	Dorke SR	Improved variety
34	Dadaba	Improved variety
35	Cidaba	Improved variety

Appendix 2: Study methods and protocols

Appendix 2.1: Geographical aspects and crop demography

Universität Bremen

Appendix 2.1.1: Mapping of small-scale maize (Zea mays) farms in peri-urban areas of Accra, Ghana.

Investigator:					Data sheet no:		Date:		Phase:	Name of community:			
		Frequency					Field location						
Field No.	Culti-var	a	b	c	Phen-ology	Vita-lity	GPS No.	GPS readings	Height (cm)	Farm Type	Sample No.	Photo No.	Additional remarks (e.g. damage, pests, etc.)

Universität Bremen

Appendix 2.1.2: Mapping of field corners of small-scale maize farms within peri-urban areas of Accra, (Ghana).

Investigator:				Community:			Date:		Data sheet No.:	
Field No.	Field corners/ outlines (GPS readings)									
	1		2		3		4		5	

Appendix 2.1.3: Protocol for estimation of temporal heterogeneity of the flowering phase of maize populations within a 1km2 area of peri-urban suburbs of Accra (Ghana) A Phase 2 investigation.

Field No.	Cultivar	Location replicates	Fl. start Plant 1	Fl. start Plant 2	Fl. end Plant 1	Fl. end Plant 2	Fl. dur Plant 1	Fl. dur Plant 2	Pheno-logy	GPS No.	GPS read-ings		Height (cm)	Farm type	Photo No.	General Remarks
	C1															
	C2															
	C3															
	C4															
	E1															
	E2															
	E3															
	E4															
	CTR															

Note: C1-C4: refers to the field corners where sampling was done; E1- E4: refers to the field edges; CTR: refers to the centre of the field.
Fl dur: Flowering duration

Appendix 2.2: Questionnaire survey on the socio-cultural and crop management conditions of small-scale maize (Zea mays) cultivation within peri-urban areas of Accra, Ghana.

Universität Bremen

This survey is a major social component of a multi-disciplinary study in partial fulfillment of a PhD thesis at the Center for Environmental Research & Technology (UFT), University of Bremen, Germany. Responses gathered will be further useful as a baseline contribution for policy-making and the improvement of small-scale agriculture in peri-urban and rural development. Thus, we rely on your cooperation to gain knowledge on crop management issues of small-scale maize cultivation practices- varieties grown, crop history, rotation measures, seed origins and criteria for seed selection or preferences. Furthermore crop production issues- time of planting, weed management, harvest, storage requirements and related constraints- e.g. pest infestation will be recorded. Sources of agricultural information and dissemination would also be documented in this exercise.

A) General information:

Date (dd/mm/yy):	S/n:
Name of investigator:	House No.:
Name of community:	Farmland location on map: (incl. GPS):

B) Respondent's profile:

Name:	Age:
Marital status: M [ ], S [ ] W [ ]	Nr. Children [ ]
Sex: M [ ], F [ ]	Place of origin:
Educational status:	Permanent job:
Part-time job(s):	Per capita monthly income:
Period of stay in the community:	Language skills:

C) General agricultural practice/ prerequisites crops currently cultivated are:

Classification of farm household types/Land tenure arrangements-
Private land [ ]; Leasehold [ ] (please specify years): ________Public land/Usufruct [ ] Other [ ] (please specify): _______________

Number fields cultivated________; Farm type: Subsistence plots/home gardens (for family supply) [ ]; Commercial plots/fields (for market supply); [ ] Labour requirements- Hired [ ]; or Family labour [ ]; of how many persons?__________
Who does what? (time spent on which activities) Men, women, children

Farming (involvement) since: 19____ / 20____

Land use history of the plot as far as it can be reconstructed - how far back? What was done at the plot at what time? This should be structured according to previous crop rotations, previous cultivation measures (please develop a timeline of activities below):

Investment values per planting season (tick category and state amount involved in ¢):
Land [ ]_______, Tillage [ ]_________, Seeds (specify quantity) [ ]_____________
Agro-chemicals [ ]_____________, Labour [ ]_____________,
Extension services [ ]____________, Machinery [ ] _______________
Income per planting season (tick category and state amount involved in ¢): Harvestable yields: Subsistence [ ] _______________ Commercial [ ] _______________

D) Socio-cultural factors/ farm management issues (site-specific)
(i) Time of planting/ harvest (current plot investigated and sampled)

Time of planting (date/week/how far back?):

Time of harvest:

When was the farm weeded? (Date/week/how far back?):

(ii) Seed origins/ acquisition and preference:
- Which varieties of maize do you plant?

- From where do you obtain these seeds? Other farmers [ ], Previous harvests [ ], Extension agents [ ], Gifts [ ], Formal seed system/ shops [ ],

If from the latter, Name of shop/ agency:_________________________

Please specify: Quantities obtained/ planting season_______ (e.g. No. of cobs or Kg);

- Are the planted varieties: traditional (landraces) [ ] or improved varieties [ ]

-What are the reasons for your particular choice of seeds above?
High yielding [ ], Stable varieties [ ], Insect resistant [ ], Low input costs [ ], Early maturing [ ], Gifts [ ]

- If seeds have been bought - are hybrids eventually re-planted? Yes [ ], No [ ].
Please explain your choice:

(iii) Tillage and weed management systems:
Weed control measures include:

Tillage [ ] Equipment used are:

Non-tillage [ ] Specify measure e.g. herbicide application and brands used are:

What is the application procedure:

Devices and protection measures (if any…):

(iv) Soil fertility management:
-Frequently employed is Green manure [ ], Compost [ ], Mulch [ ],
Inorganic fertilizer [ ], Other [ ], None of these [ ]
(Please give reasons for your choice(s) above):

(v) Limiting factors to productivity: Rainfall variability [ ], insect pests/aphids, stem borers, termites, ants, root worms, others [ ] (which?) Wild Animals [ ] (which? Monkeys, birds? domestic animals (goats)?..............................,
Nutrient availability [ ], Diseases fungi, viruses, which[ ]? ....,
Erosion/ soil conservation [ ], Post-harvest losses [ ], Others [ ] (please specify) (Please give typical examples):

Which pests most affect productivity (priority pests): Stem borers [ ], Root feeders (e.g. termites) [ ], Storage pests (e.g. weevils, moths), Parasitic weed (e.g. Striga hermonthica) [ ], Other [ ]..................................),
Have you encountered diseases within your farm? No [ ], yes [ ]; If yes, in which category would you classify the disease: Maize steak virus [ ], Fungal pathogen [ ], Other [ ] (Please specify________________________).

How are these overcome?

(vi) Harvesting & post harvest issues: - Harvest techniques employed are:

-What quantity (incl. %) of harvested quantities are stored for re-planting in a next season?

-Which criteria are used in seed (kernels) selection?

-What are the storage requirements?

-Storage facilities: Silos [ ], Traditional shelves/ barns [ ], Others [ ]; Shelf life: _____________(months/years),

Which of the two would you value: [ ] more stable, lower-yielding varieties from local, inexpensive seed sources requiring little inputs [ ] initially high yielding from the formal seed system, requiring higher inputs (fertilizers, plant protection), more expensive with declining yields (an assumption as may be for e.g. with Bt varieties to which pests develop resistance over time)

(vii) Information dissemination and management:
-In which way is agricultural information disseminated:
Farmer-Farmers [ ], Extension service depts.[ ], Media [ ], Research project [ ], Others (specify):

Agricultural extension requirements in your case? no [ ], yes [ ].

If yes, any data/reference on activities, eventual application and dissemination (where, how many farmers involved):

-Are there existing forms of informal or volunteer co-operation e.g. neighbourhood help during peak work occasions no [ ], yes [ ]

-If yes, Please specify activities:

Appendix 3: Documentation of reagents, materials and instruments used

Table A: List of reagents.

Reagent	Company
Absolute ethanol	Riedel-de Haen, Seelze
Agarose NEEO Ultra-Qualität	Roth, Karlsruhe
Boric acid	Fluka, Buchs, Switzerland
BSA, 10 mg/ml, 5 mg	Fermentas, St. Leon-Rot
Chelex® 100 resin	BioRad, Hercules, USA
DNeasy® Plant Mini Kit (50)	Qiagen, Hilden
dNTP Mix, 10 mM each, 1 ml	Fermentas, St. Leon-Rot
Ethidium bromide, 0.4 %	Serva, Heidelberg
Extran® AP12	Merck, Darmstadt
Magnesiumchloride, 25 mM, 1.5 ml	Fermentas, St. Leon-Rot
6x Orange Loading Dye, 1 ml	Fermentas, St. Leon-Rot
O´RangeRuler 20bp DNA Ladder, 0.1 µg/µl, 500 µg	Fermentas, St. Leon-Rot
PCR buffer, 10x, 1.5 ml, with 15 mM MgCl2	GeneCraft, Lüdinghausen
TaqDNA Polymerase, 5 U/µl, 500 U	GeneCraft, Lüdinghausen
Titriplex® (Na2EDTA)	Merck, Darmstadt
Trizma®Base	Roth, Karlsruhe

Table B: List of materials.

Material	**Company**
Beaker	Schott Duran, Mainz
Cryo boxes	Omnilab, Bremen
Duran flasks with caps, several sizes	Schott Duran, Mainz
Erlenmeyer flasks, several sizes	Schott Duran, Mainz
Gel electrophoresis chamber, slides and combs	BioRad, Hercules, USA
Measuring cylinder	Schott Duran, Mainz
Pipette tips, several sizes	Eppendorf, Hamburg
Pipettes, several sizes	Eppendorf, Hamburg
Reaction tubes, several sizes	Eppendorf, Hamburg

Table C: List of instruments.

Instrument	**Company**
Centrifuge mini spin	Eppendorf, Hamburg
Heating cabinet	Memmert, Schwabach
Heating cabinet	Heraeus Instruments, Düsseldorf
Mixer Vortex-Genie 2	Scientific Industries, Bohemia, USA
NanoDrop 1000 Spectrophotometre	PeqLab, Erlangen
Power Supply Power Pac 3000	BioRad, Hercules, USA
Shaker TH 15	Edmund Bühler, Tübingen
Thermocycler iCycler	BioRad, Hercules, USA
Thermomixer compact	Eppendorf, Hamburg
Ultrapure water purification system	SG Reinstwasser, Barsbüttel
UV-transilluminator	Vilber Lourmat, Marne-la-Vallée, France
Weighing machine	Sartorius Laboratory, Göttingen

Description of prepared solution

3 % Agarose gel: 10.5 g agarose was dissolved in 200 ml 1x TBE buffer. After complete dissolution the mixure was filled up with 1x TBE buffer to a final volume of 350 ml.

TBE (tris-borate) buffer: 10x TBE buffer = 108 g Trizma® Base, 55 g boric acid and 7.45 g Titriplex were dissolved in 700 ml ddH_2O. After complete dissolution the buffer was filled up with ddH_2O to a final volume of 1000 ml. 1x TBE buffer = 10x TBE buffer was diluted in a ratio of 1:10 with ddH_2O.

0.5 mg/ml ethidium bromide solution: 62.5 µl 0.4 % ethidium bromide stock solution was mixed with 500 ml 1x TBE buffer.

70 % ethanol: 70 ml ethanol absolute was mixed with 30 ml ddH_2O.

3 % extran solution: 30 g extran was dissolved in 800 ml ddH_2O. After complete dissolution the mixture was filled up with ddH_2O to a final volume of 1000 ml.

5 % Chelex: 5 g Chelex was dissolved in 80 ml ddH_2O. After complete dissolution the mixture was filled up with ddH_2O to a final volume of 100 ml.

PCR conditions and thermal cycle profiles

Table D: PCR conditions of the primer pair phi99100.

Reagent	**Concentration of stock solution**	**Concentration in reaction solution**	**µl per tube**
ddH_2O	- - -	- - -	27.5
PCR buffer with $MgCl_2$	10x / 15 mM $MgCl_2$	1x /1.5 mM $MgCl_2$	5
$MgCl_2$	25 mM	1.25 mM	2.5
dNTPs	10 mM	0.4 mM	2
Primer pair **phi99100**	10 µM	0.5 µM	2.5
BSA	10 mg/ml	0.6 mg/ml	3
DNA extract	- - -	- - -	4
Taq polymerase	5 U/µl	0.1 U/µl	1
Final volume			**50**

Table E: PCR conditions of the primer pairs phi1084111 and phi031.

Reagent	Concentration of stock solution	Concentration in reaction solution	µl per tube
ddH$_2$O	- - -	- - -	26.3
PCR buffer with MgCl$_2$	10x / 15 mM MgCl$_2$	1x /1.5 mM MgCl$_2$	5
MgCl$_2$	25 mM	1.25 mM	2.5
dNTPs	10 mM	0.4 mM	2
Primer pairs **phi1084111** and **phi031**	10 µM	0.6 µM	3
BSA	10 mg/ml	0.6 mg/ml	3
DNA extract	- - -	- - -	4
Taq polymerase	5 U/µl	0.12 U/µl	1.2
Final volume			**50**

Table F: PCR conditions of the primer pair phi331888.

Reagent	Concentration of stock solution	Concentration in reaction solution	µl per tube
ddH$_2$O	- - -	- - -	25.3
PCR buffer with MgCl$_2$	10x / 15 mM MgCl$_2$	1x /1.5 mM MgCl$_2$	5
MgCl$_2$	25 mM	1.25 mM	2.5
dNTPs	10 mM	0.4 mM	2
Primer pairs **phi331888**	10 µM	0.7 µM	3,5
BSA	10 mg/ml	0.6 mg/ml	3
DNA extract	- - -	- - -	4
Taq polymerase	5 U/µl	0.12 U/µl	1.2
Final volume			**50**

Table G: Thermal cycle profile of all conducted PCRs

Step	Temperature	Time	Cycles
Initial denaturation	94° C	5 minutes	1
Denaturation	94° C	30 seconds	30
Annealing	40° C	60 seconds	
Elongation	72° C	60 seconds	
Final elongation	72° C	5 minutes	1

Acknowledgements

I thank God for bringing me to the conclusion of my PhD studies. Several people have been instrumental leading to the finalization of this work. I am highly indebted to PD Dr. Broder Breckling, my first supervisor for his immense scientific guidance, supervision and mutual support in the advancement of this topic. I am very thankful to Broder for introducing me to one of the significant environmental and current scientific issues of interest to mankind. Your support was unflinching even in the most turbulent and frustrating moments of this study. I am grateful. I am extremely grateful to Prof. Dr. Juliane Filser, second supervisor and head of Dept. of Theoretical and General Ecology for accepting me as a doctoral student in her work group at the UFT. Juliane, your scientific support including provision of relevant recommendations for acquisition of funds and discussions imposed an important contribution. I thank Prof. Dr. Wolfgang Heyser, former Director of the UFT, and PD Dr. Karin Mathes, Vice-President of the Bremen Parliament for examining the thesis. I am also grateful to my local advisor Mr. Alex Owusu-Biney, Coordinator of the African Biosafey Programmes (UNEP, Nairobi) who provided local institutional affiliation at the Biosafety Secretariat in Accra and discussed this topic. I express gratitude also to the Director of BNARI (Biotechnology and Nuclear Agriculture Research Institute), Prof. Josephine Nketia for accepting me as a guest student at her institute. I am indeed very thankful to the following institutions for financial awards granted in the final steps of this work in Bremen:

- The DAAD Prize for Scientific Merits and Social Engagement at the University of Bremen -2007
- DAAD-STIBET PhD Teaching and Research Scholarship at the University of Bremen
- KAAD (Katholischer Akademischer Austauschdienst), Bonn
- KHG (Katholischer Hochschule Gemeinde), Bremen

My special gratitude goes to Heidi Wolter, Dr.Ulrich Burkhardt and Iris Burfeindt for technical assistance with the molecular experiments in the UFT laboratories. I am thankful to Dr. Marilyn Warburton, CIMMYT (Mexico) for insightful discussion on the molecular aspects. I wish to use this occasion to thank Christian Aden of the Dept. of Landscape Ecology of the University of Vechta for support with the GIS applications. My gratitude also goes to PD Dr. Hauke Reuter for assistance with the application of the maize model developed in the context of the BMBF-GeneRisk Project at the UFT. On this occasion, I would also like to extend my immense appreciation to my field research assistants- Eric Blabou, Emmanuel Kuchula, and William Addai for their contribution in data collection and compilation, certainly was a tedious task. I thank immensely my guest family- Klaus and Suzana Muthreich, who largely secured my stay in Bremen in the most difficult periods of my studies in Bremen. But for your support, this study would not have been brought to a successful conclusion. I wish to thank my dear family both big and small for their constant source of inspiration. To Cynthia and Sefa, I am particularly grateful for allowing to use your time for this work and for your enduring patience. I thank my brother Peter Aheto for all the rounds he did in respect of this work. I wish to thank all my colleagues in the working group AG Filser for the congenial atmosphere and support you offered during this period. I would like to acknowledge friends like Hendra Kesuma for the quality time we spent together in Bremen. I am most indebted to Dr. Richard Verhoeven of the University of Bremen for providing technical support with the final text formatting and for addressing pertinent layout issues with respect to this publication. Finally, my heartfelt gratitude goes to the many farmers and their families who allowed us gather data on their farms and took the time to participate in this survey. Stay blessed.

Theorie in der Ökologie

Herausgegeben von Broder Breckling

www.peterlang.de